AJ Sadler

Mathematics Methods 2

Student Book

Unit 2

Australia • Brazil • Japan • Korea • Mexico • Singapore • Spain • United Kingdom • United States

Mathematics Methods Unit 2
1st Edition
A.J. Sadler

Publishing editor: Robert Yen
Project editor: Alan Stewart
Cover design: Chris Starr (MakeWork)
Text designers: Sarah Anderson, Nicole Melbourne, Danielle Maccarone
Permissions researcher: Jan Calderwood
Answer checker: George Dimitriadis
Production controller: Erin Dowling
Typeset by: Cenveo Publisher Services

National Library of Australia Cataloguing-in-Publication Data
Sadler, A.J., author.
Mathematics methods : unit 2 / A.J. Sadler.

1st revised edition
9780170390408 (paperback)
Includes index.
For secondary school age.

Mathematics--Study and teaching (Secondary)--Australia.
Mathematics--Textbooks.

510.712

Cengage Learning Australia
Level 7, 80 Dorcas Street
South Melbourne, Victoria Australia 3205

Cengage Learning New Zealand
Unit 4B Rosedale Office Park
331 Rosedale Road, Albany, North Shore 0632, NZ

For learning solutions, visit **cengage.com.au**

Printed in China by 1010 Printing International Limited.
9 10 11 12 25 24 23

PREFACE

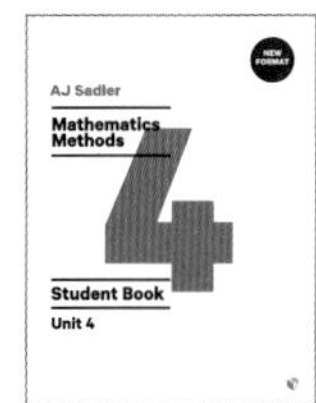

This text targets Unit Two of the West Australian course *Mathematics Methods*, a course that is organised into four units, units one and two for year eleven and units three and four for year twelve.

The West Australian course, *Mathematics Methods*, is based on the Australian Curriculum Senior Secondary course *Mathematical Methods*. Apart from small changes to wording, the unit twos of these courses are closely aligned but do differ in that the Australian Curriculum course, when considering differentiation, goes beyond polynomial functions and considers other linear combinations of power functions, and when considering antiderivatives, includes solving problems involving motion in a straight line. At the time of writing, these two aspects are not included in unit two of the West Australian course. I have found it appropriate to include the first of these items in this book, in Chapter 5, and in later miscellaneous exercises, as it allows students to realise that the rule for differentiating ax^n, with respect to x, can be applied for n negative and/or fractional. To maintain alignment with Australian Curriculum I include the second item in Chapter 8. West Australian users of this book can decide whether to include or omit these aspects.

In a few other places I have found it appropriate to go a little outside the confines of the syllabus for the unit. In this regard I include consideration of infinite geometric series, I include a few optimisation questions involving functions that are not polynomials (as an extension exercise in Chapter 6) and on the basis that students are likely to encounter the integral sign on their calculators, I include this in the chapter on antidifferentiation.

The book contains text, examples and exercises containing many carefully graded questions. A student who studies the appropriate text and relevant examples should make good progress with the exercise that follows.

The book commences with a section entitled **Preliminary work**. This section briefly outlines work of particular relevance to this unit that students should either already have some familiarity with from the mathematics studied in earlier years, or for which the brief outline included in the section may be sufficient to bring the understanding of the concept up to the necessary level.

As students progress through the book they will encounter questions involving this preliminary work in the **Miscellaneous exercises** that feature at the end of each chapter. These miscellaneous exercises also include questions involving work from preceding chapters to encourage the continual revision needed throughout the unit.

Some chapters commence with a '**Situation**' or two for students to consider, either individually or as a group. In this way students are encouraged to think and discuss a situation, which they are able to tackle using their existing knowledge, but which acts as a fore-runner and stimulus for the ideas that follow. Students should be encouraged to discuss their solutions and answers to these situations and perhaps to present their method of solution to others. For this reason answers to these situations are generally not included in the book.

Alan Sadler

ISBN 9780170390408

CONTENTS

ISBN 9780170390408

ISBN 9780170390408

IMPORTANT NOTE

This series of texts has been written based on my interpretation of the appropriate *Mathematics Methods* syllabus documents as they stand at the time of writing. It is likely that as time progresses some points of interpretation will become clarified and perhaps even some changes could be made to the original syllabus. I urge teachers of the *Mathematics Methods* course, and students following the course, to check with the appropriate curriculum authority to make themselves aware of the latest version of the syllabus current at the time they are studying the course.

Acknowledgements

As with all of my previous books I am again indebted to my wife, Rosemary, for her assistance, encouragement and help at every stage.

To my three beautiful daughters, Rosalyn, Jennifer and Donelle, thank you for the continued understanding you show when I am 'still doing sums' and for the love and belief you show.

Alan Sadler

ISBN 9780170390408

PRELIMINARY WORK

This book assumes that you are already familiar with a number of mathematical ideas from your mathematical studies in earlier years.

This section outlines the ideas which are of particular relevance to Unit Two of the **Mathematics Methods** course and for which familiarity will be assumed, or for which the brief explanation given here may be sufficient to bring your understanding of the concept up to the necessary level.

Read this 'Preliminary work' section and if anything is not familiar to you, and you don't understand the brief mention or explanation given here, you may need to do some further reading to bring your understanding of those concepts up to an appropriate level for this unit. (If you do understand the work but feel somewhat 'rusty' with regards to applying the ideas some of the chapters afford further opportunities for revision as do some of the questions in the miscellaneous exercises at the end of chapters.)

Note especially the last inclusion in this section which involves factorising expressions of the form $a^n - b^n$. Other than that of $a^2 - b^2$ the reader is probably not familiar with the factorisations given here, but by reading the brief explanation given the reader should bring their understanding of the concept up to the required understanding for its later use in this text.

- Chapters in this book will continue some of the topics from this Preliminary work by building on the assumed familiarity with the work.
- The **miscellaneous exercises** that feature at the end of each chapter may include questions requiring an understanding of the topics briefly explained here.

Number

It is assumed that you are already familiar with counting numbers, whole numbers, integers, factors, multiples, prime numbers, composite numbers, square numbers, negative numbers, fractions, decimals, the rule of order, percentages, the square root and the cube root of a number, powers of numbers (including zero and negative powers), and can use this familiarity appropriately. An ability to simplify simple expressions involving square roots is also assumed.

e.g.

$$\begin{aligned}\sqrt{8} &= \sqrt{4\times 2}\\ &= 2\sqrt{2}\end{aligned}$$

$$\begin{aligned}\sqrt{27}+\sqrt{75} &= \sqrt{9\times 3}+\sqrt{25\times 3}\\ &= 3\sqrt{3}+5\sqrt{3}\\ &= 8\sqrt{3}\end{aligned}$$

An understanding of numbers expressed in *standard form* or *scientific notation*, e.g. writing 260 000 in the form 2.6×10^5 or writing 0.0015 in the form 1.5×10^{-3}, is also assumed.

Percentages

It is assumed you are familiar and comfortable with the use of percentages and in particular their use in the concepts of **simple interest** and **compound interest**.

ISBN 9780170390408

Rounding

Answers to some calculations may need rounding to a suitable or specified accuracy. For example, if we were to cut a 5 metre length of string into seven equal pieces it would be unwise to claim that each piece would be of length 0.714 285 714 3 metres, the answer a calculator might give when asked to calculate 5 ÷ 7. Not only is this answer too accurate for the task involved, it is also ludicrous to claim such accuracy when the information used to calculate it, i.e. the length being 5 metres, would not have been measured to this accuracy itself. Instead we might round the answer to perhaps 2 decimal places. I.e. 0.71 metres.

In some cases the situation may make *truncating* more appropriate than rounding. Suppose for example we have \$10 and wish to buy as many chocolate bars costing \$2.15 each as possible. Whilst \$10 ÷ \$2.15 is 4.65 if we round to two decimal places, 4.7 if we round to one decimal place and 5 if we round to the nearest integer, a more appropriate answer is obtained by truncating to 4 as that is the number of chocolate bars we would be able to buy with our \$10 (and we would have \$1.40 change). If we truncate to an integer we discard the decimal part entirely.

Suppose the estimated cost to a company of manufacturing 18 models of a new machine is \$375 000. Dividing this amount by 18 gives an estimated cost of \$20 833·33 each (nearest cent).

However, this is again likely to be too accurate for the situation and we might instead round to the nearest \$100, in this case \$20 800. Alternatively we could say that we have rounded to three *significant figures*. To round to a number of significant figures we count that number of digits and then use the next digit to apply our usual rounding rules.

For example	526 086.9565	is	500 000	to 1 significant figure,
		is	530 000	to 2 significant figures,
		is	526 000	to 3 significant figures,
		is	526 100	to 4 significant figures, etc.

For very small numbers we do not count any initial zeros as significant figures.

For example	0.000 310 486 2	is	0.0003	to 1 significant figure,
			0.000 31	to 2 significant figures,
			0.000 310	to 3 significant figures,
			0.000 310 5	to 4 significant figures, etc.

Indices

Whilst it is anticipated that you are familiar with the idea of

- raising a number to some *power*,
- zero and negative integers as powers,
- fractions as powers,

and that you may well be aware of some of the *index laws*, these ideas will be re-visited in the first chapter of this text.

ISBN 9780170390408

Function

It is assumed that the reader is familiar with the idea of a **function** being a rule that associates with each element in a set S a unique element from a set T. The set S is called the **domain** of the function, the set T is the **codomain** of the function and those elements of T that the function maps elements of S *onto* form the **range** of the function. If the domain of a function is not specifically stated then we assume it to be the set of all real numbers for which the function is defined. This is the **natural domain** of the function. If the function $f(x)$ maps the element a from the domain, onto the element b, of the range, we write $f(a) = b$.

The requirement that a function takes an element of the domain and maps it onto one, and only one, element of the range means that the graph of a function will pass the *vertical line test*. I.e. if a vertical line is moved from the left end of the x-axis to the right end of the x-axis it will not cut the graph at any more than one place at a time.

The reader should be familiar with the concept of polynomial functions in general and be especially familiar with linear, quadratic and cubic polynomials. This familiarity should include an understanding of the key features of the graphs of these particular types of functions, and of the reciprocal function. These key graphical features including intercepts with the axes, turning points, points of inflection, asymptotes, concavity, symmetry and the behaviour of the function as $x \to \pm\infty$.

Linear, quadratic and reciprocal functions in particular

Linear functions have:

- *Equations* of the form $y = mx + c$.
- *Graphs* that are straight lines with gradient m and cutting the y-axis at $(0, c)$.
- *Tables of values* that have a *constant first difference* pattern.

Quadratic functions have

- *Equations* of the form: $y = ax^2 + bx + c$, sometimes written as $y = a(x - p)^2 + q$ or $y = a(x - d)(x - e)$ each form allowing various key features of the graph of the quadratic function to be readily determined.
- *Graphs* that are parabolic in shape with either a 'hill', **a maximum**, or a 'valley', **a minimum**.
- *Tables of values* with a *constant second difference* pattern.

Reciprocal functions have

- *Equations* of the form $y = \frac{k}{x}, x \neq 0$.
- *Graphs* that are said to be **hyperbolic** in shape with the x- and y-axes as **asymptotes** to the curve.
- *Table of values* that have a constant product (equal to k).

Transformations

The reader should be familiar with how the graphs of

$y = f(x) + k$,

$y = f(x - k)$,

$y = af(x)$,

and $y = f(ax)$

relate to that of $y = f(x)$.

Adding k to the right hand side

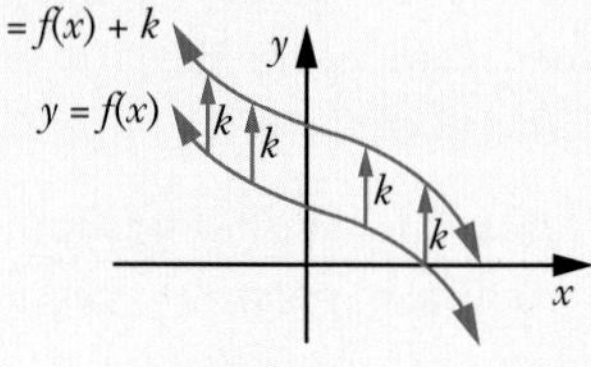

The graph of $y = f(x) + k$ will be that of $y = f(x)$ translated k units vertically upwards.

Thus if k is negative the translation will be vertically downwards.

'Replacing x by (x – k).'

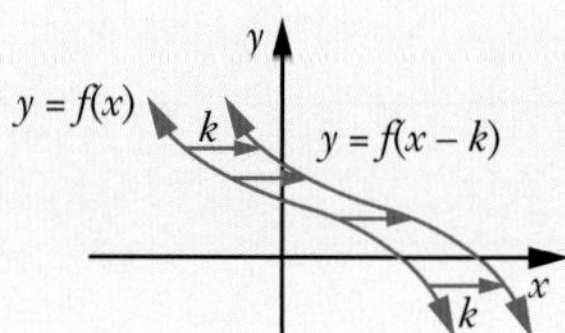

The graph of $y = f(x - k)$ will be that of $y = f(x)$ translated k units to the right.

Thus if k is negative the translation will be to the left.

'Multiplying the right hand side by –1.'

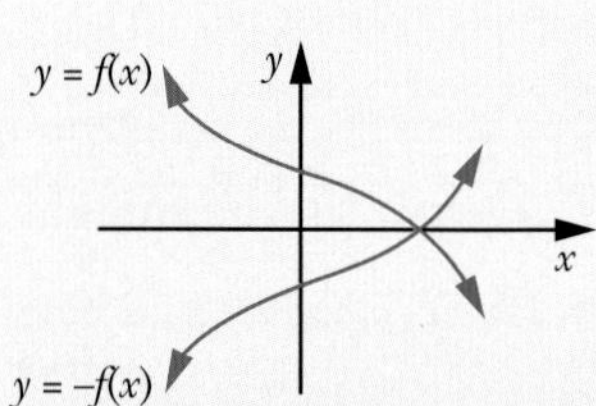

The graph of $y = -f(x)$ will be that of $y = f(x)$ reflected in the x-axis.

'Replacing x by –x.'

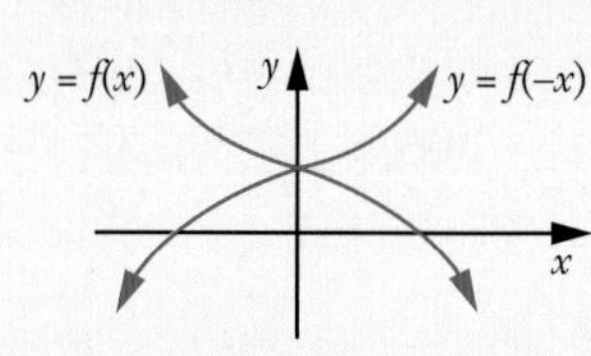

The graph of $y = f(-x)$ will be that of $y = f(x)$ reflected in the y-axis.

'Multiplying the right hand side by a.'

The graph of $y = af(x)$ will be that of $y = f(x)$ dilated parallel to the y-axis with scale factor a. A point that is q units above the x-axis will be moved vertically to a point that is aq units above the x-axis. Points on the x-axis will not move.

If $a > 1$ the effect will be to stretch $y = f(x)$ vertically and if $0 < a < 1$ the effect will be to compress $y = f(x)$ vertically.

'Replacing x by ax.'

The graph of $y = f(ax)$ will be that of $y = f(x)$ dilated parallel to the x-axis with scale factor $\frac{1}{a}$. A point that is p units from the y-axis will be moved horizontally to a point that is $\frac{p}{a}$ units from the y-axis. Points on the y-axis will not move.

If $a > 1$ the effect will be to compress $y = f(x)$ horizontally and if $0 < a < 1$ the effect will be to stretch $y = f(x)$ horizontally.

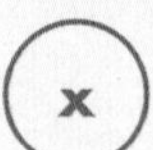

ISBN 9780170390408

Equations

The reader should be able to solve linear equations, linear simultaneous equations and be familiar with factorisation, completing the square and the formula approach for solving quadratic equations. Familiarity with the ability of some calculators to solve equations is also assumed.

Coordinates

Whilst the reader may well be familiar with finding the length of a line joining two points and with determining the midpoint of a line joining two points, the most significant piece of information for this unit is the **gradient** of the line joining two points:

If a line passes through two points, A and B, then the gradient of the line is:

$$\frac{\text{the change in the } y\text{-coordinate in going from } A \text{ to } B}{\text{the change in the } x\text{-coordinate in going from } A \text{ to } B}$$

Thus the gradient of the straight line through $A(x_1, y_1)$ and $B(x_2, y_2) = \dfrac{y_2 - y_1}{x_2 - x_1}$.

Note that in the previous formula, whilst $\dfrac{y_1 - y_2}{x_1 - x_2}$ would also give the correct answer, $\dfrac{y_1 - y_2}{x_2 - x_1}$ and $\dfrac{y_2 - y_1}{x_1 - x_2}$ would not.

Also remember that if two lines are perpendicular then the product of their gradients is –1. For example lines with gradients of 2 and $-\dfrac{1}{2}$ are perpendicular.

^nC_r

We use the notation ^nC_r for the number of combinations of r different objects taken from a set containing n different objects.

There are ^nC_r combinations of r objects chosen from n different objects where $^nC_r = \dfrac{n!}{(n-r)!r!}$.

Thus the number of combinations of three objects chosen from five different objects will be

$$\begin{aligned} ^5C_3 &= \frac{5!}{(5-3)!3!} \\ &= \frac{5!}{2!3!} \\ &= \frac{5\times 4}{2\times 1} \\ &= 10 \end{aligned}$$

ISBN 9780170390408

Expanding $(a + b)^n$

The expansion of $(a + b)^n$ will be of the form

$$k_0a^n + k_1a^{n-1}b^1 + k_2a^{n-2}b^2 + k_3a^{n-3}b^3 + \ldots.. + k_nb^n$$

The first term is formed when we multiply the a from each bracket together to form a^n.

For the second term we must choose one of the n brackets to supply the b and the others will then all supply an a. This can be done in ${}^nC_1$ ways.

For the third term we must choose two of the n brackets each to supply b and the others will then each supply an a. This can be done in ${}^nC_2$ ways.

Continuing with this approach leads to the binomial expansion:

$$(a + b)^n = a^n + {}^nC_1a^{n-1}b^1 + {}^nC_2a^{n-2}b^2 + {}^nC_3a^{n-3}b^3 + \ldots + {}^nC_na^0b^n$$

this formula giving the same expansion for $(a + b)^2$, $(a + b)^3$, $(a + b)^4$ etc. as we would obtain by multiplying out the brackets, or by obtaining the coefficients k_0, k_1, k_2 etc. from the appropriate line of Pascal's triangle.

Factorising $a^n - b^n$

You should be familiar with the fact that $a^2 - b^2 = (a - b)(a + b)$.

However, probably not so familiar are the following facts:

$$a^3 - b^3 = (a - b)(a^2 + ab + b^2)$$

$$a^4 - b^4 = (a - b)(a^3 + a^2b + ab^2 + b^3)$$

$$a^5 - b^5 = (a - b)(a^4 + a^3b + a^2b^2 + ab^3 + b^4)$$

$$a^6 - b^6 = (a - b)(a^5 + a^4b + a^3b^2 + a^2b^3 + ab^4 + b^5)$$

Check the validity of each of the above by expanding the right hand side in each case.

To generalise:

$$a^n - b^n = (a - b)(a^{n-1} + a^{n-2}b + a^{n-3}b^2 + a^{n-4}b^3 + \ldots ab^{n-2} + b^{n-1})$$

ISBN 9780170390408

Indices

- Revision of powers, or indices
- Solving equations involving indices
- What if we cannot solve the equation mentally or algebraically?
- Miscellaneous exercise one

Situation One

A tale of two students (I)

Two students are asked to work out $5^{17} \div 5^{15}$ without the help of a calculator.

One of the students starts to work out the powers of 5:

$$5^2 = 5 \times 5 = 25$$
$$5^3 = 5 \times 5 \times 5 = 125$$
$$5^4 = 5 \times 5 \times 5 \times 5 = 625$$
$$5^5 = 5 \times 5 \times 5 \times 5 \times 5 = 3125, \text{ etc.}$$

The other student simply looks at $5^{17} \div 5^{15}$ and says:

'The answer must be 25'.

How is the second student able to work out the answer so quickly?

Situation Two

A tale of two students (II)

Two students are asked to work out how much will be in a savings account after 25 years if \$5000 is invested in the account at the beginning of the 25 years, interest is added at 10% per year and interest added in one year itself earns interest in subsequent years (i.e. *compound interest* is involved).

One of the students starts working on a year by year basis as follows:

Amount in account during year 1 = \$5000

At the end of year 1, interest earned = 10% of \$5000 (i.e. \$500)

Amount in account during year 2 = \$5500 (= \$5000 + \$500)

At the end of year 2, interest earned = 10% of \$5500 (i.e. \$550)

Amount in account during year 3 = \$6050 (= \$5500 + \$550)

At the end of year 3, interest earned = 10% of \$6050 (i.e. \$605)

Amount in account during year 4 = \$6655, etc

The other student does a quick calculation using a calculator and said:

'After 25 years the account will be worth \$54 173.53'.

How was this second student able to determine the answer so quickly?

Situation Three

A certain sum of money, $\$P$, is invested at 8% interest compounded annually.

How many years will it take to become $\$(4P)$?

What if the interest rate had been 12% rather than 8%?

Shutterstock.com/Robyn Mackenzie

ISBN 9780170390408

Review of index laws

Index laws review

Indices squaresaw

Revision of powers, or indices

The first situation on the previous page involved a number being raised to some *power*, or *exponent*, and the other two situations could be solved using this idea. As was mentioned in the *Preliminary work* section at the beginning of this book it is anticipated that you are familiar with this idea and are also familiar with *zero and negative integers as powers* and that you may be aware of some of the *index laws*.

Note

In an expression like 2^3, the number 3 is the *index* and the number 2 is the *base*. The index shows the *power* to which the base is raised.

Read through the following to revise, and possibly extend, your understanding of these concepts and then work through the exercise that follows to practise the application of the ideas.

Notice that

$$a^2 \times a^3 = (a \times a) \times (a \times a \times a) = a^5 \quad (= a^{2+3})$$
$$a^3 \times a^5 = (a \times a \times a) \times (a \times a \times a \times a \times a) = a^8 \quad (= a^{3+5})$$
$$a^2 \times a^7 = (a \times a) \times (a \times a \times a \times a \times a \times a \times a) = a^9 \quad (= a^{2+7})$$

To generalise:

$$a^n \times a^m = a^{n+m}$$

Notice that

$$a^5 \div a^3 = \frac{\cancel{a \times a \times a} \times a \times a}{\cancel{a \times a \times a}} = a^2 \quad (= a^{5-3})$$

$$a^7 \div a^4 = \frac{\cancel{a \times a \times a \times a} \times a \times a \times a}{\cancel{a \times a \times a \times a}} = a^3 \quad (= a^{7-4})$$

$$a^8 \div a^3 = \frac{\cancel{a \times a \times a} \times a \times a \times a \times a \times a}{\cancel{a \times a \times a}} = a^5 \quad (= a^{8-3})$$

To generalise:

$$a^n \div a^m = a^{n-m}$$

From the above rule it follows that

$$a^5 \div a^5 = a^0$$
$$a^7 \div a^7 = a^0$$
$$a^{12} \div a^{12} = a^0$$

However, $a^5 \div a^5$, $a^7 \div a^7$ and $a^{12} \div a^{12}$ each involve something divided by itself, which must equal 1 (provided the 'something' does not equal zero).

Hence

$$a^0 = 1$$

Again using the fact that $a^n \div a^m = a^{n-m}$, it follows that

$$a^0 \div a^n = a^{-n}$$

But

$$a^0 \div a^n = 1 \div a^n = \frac{1}{a^n}$$

Hence

$$a^{-n} = \frac{1}{a^n}$$

ISBN 9780170390408

EXAMPLE 1

Evaluate each of the following without the use of a calculator.

a $8^{19} \div 8^{17}$ **b** $3^0 + 2 \times 5^0$ **c** $2^8 \div 8$ **d** $\dfrac{6^3 \times 6^7}{6^8}$

Solution

a $8^{19} \div 8^{17} = 8^{19-17}$
$= 8^2$
$= 64$

b $3^0 + 2 \times 5^0 = 1 + 2 \times 1$
$= 1 + 2$
$= 3$

c $2^8 \div 8 = 2^8 \div 2^3$
$= 2^5$
$= 32$

d $\dfrac{6^3 \times 6^7}{6^8} = 6^{3+7-8}$
$= 6^2$
$= 36$

EXAMPLE 2

Express each of the following as a power of 5 (i.e., in the form 5^k).

a $5^8 \div 5^5$ **b** $5^7 \times 25$ **c** $\dfrac{1}{5^4}$ **d** 0.2

Solution

a $5^8 \div 5^5 = 5^{8-5}$
$= 5^3$

b $5^7 \times 25 = 5^7 \times 5^2$
$= 5^9$

c $\dfrac{1}{5^4} = 5^{-4}$

d $0.2 = \dfrac{1}{5}$
$= 5^{-1}$

Using the fact that $a^n \times a^m = a^{n+m}$, it follows that

$$a^{\frac{1}{2}} \times a^{\frac{1}{2}} = a^1$$

Thus $a^{\frac{1}{2}}$ multiplied by itself gives a. Hence $a^{\frac{1}{2}}$, or $a^{0.5}$, is the square root of a.

Similarly, $a^{\frac{1}{3}}$ is the cube root of a, $\sqrt[3]{a}$, $a^{\frac{1}{4}}$ is the fourth root of a, $\sqrt[4]{a}$,

$a^{\frac{1}{5}}$ is the fifth root of a, $\sqrt[5]{a}$, etc.

Hence

$$a^{\frac{1}{n}} = \sqrt[n]{a}$$

ISBN 9780170390408

Note: A power of $\frac{1}{2}$ represents the **positive** square root, just as the radical sign, $\sqrt{\ }$, does.

That is, if $x = \sqrt{4}$ then $x = 2$.

If $x = 4^{\frac{1}{2}}$ then $x = 2$.

But if $x^2 = 4$ then $x = \pm 2$.

Notice that

$$(a^2)^3 = (a \times a) \times (a \times a) \times (a \times a) = a^6 \quad (= a^{2 \times 3})$$
$$(a^4)^2 = (a \times a \times a \times a) \times (a \times a \times a \times a) = a^8 \quad (= a^{4 \times 2})$$
$$(a^3)^3 = (a \times a \times a) \times (a \times a \times a) \times (a \times a \times a) = a^9 \quad (= a^{3 \times 3})$$

To generalise:

$$(a^n)^m = a^{n \times m}$$

Notice that

$$(ab)^2 = (a \times b) \times (a \times b) = a^2 \times b^2$$
$$(ab)^3 = (a \times b) \times (a \times b) \times (a \times b) = a^3 \times b^3$$
$$(ab)^4 = (a \times b) \times (a \times b) \times (a \times b) \times (a \times b) = a^4 \times b^4$$

To generalise:

$$(ab)^n = a^n \times b^n$$

Notice that

$$\left(\frac{a}{b}\right)^5 = \frac{a}{b} \times \frac{a}{b} \times \frac{a}{b} \times \frac{a}{b} \times \frac{a}{b} = \frac{a^5}{b^5}$$

To generalise:

$$\left(\frac{a}{b}\right)^n = \frac{a^n}{b^n}$$

EXAMPLE 3

Evaluate each of the following without the use of a calculator.

a $16^{0.5}$ **b** $(2^6 \div 2^4)^2$ **c** $\left(3\frac{3}{8}\right)^{\frac{1}{3}}$ **d** $8^{\frac{4}{3}}$

Solution

a $16^{0.5} = \sqrt{16}$
$= 4$

b $(2^6 \div 2^4)^2 = (2^2)^2$
$= 2^4$
$= 16$

c $\left(3\frac{3}{8}\right)^{\frac{1}{3}} = \sqrt[3]{\frac{27}{8}}$
$= \frac{\sqrt[3]{27}}{\sqrt[3]{8}}$
$= \frac{3}{2}$ i.e., 1.5

d $8^{\frac{4}{3}} = (8^{\frac{1}{3}})^4$
$= 2^4$
$= 16$

ISBN 9780170390408

The previous answers can be confirmed with a calculator.

$16^{0.5}$	
	4
$(2^6 / 2^4)^2$	
	16
$\sqrt[3]{3+\frac{3}{8}}$	
	1.5
$8^{\frac{4}{3}}$	
	16

EXAMPLE 4

Simplify each of the following, expressing your answers in terms of positive indices.

a $5a^2y^3 \times 6a^4y^7$ **b** $\dfrac{25ab^9}{15a^7b^4}$ **c** $\left(a^{-7} \times a^3\right)^{\frac{1}{2}}$ **d** $\left(\dfrac{a^3b}{b^3}\right)^{-2}$

Solution

a $5a^2y^3 \times 6a^4y^7 = 30a^{2+4}y^{3+7}$
$= 30a^6y^{10}$

b $\dfrac{25ab^9}{15a^7b^4} = \dfrac{25}{15} \times a^{1-7}b^{9-4}$
$= \dfrac{5}{3} \times a^{-6}b^5$
$= \dfrac{5b^5}{3a^6}$

c $(a^{-7} \times a^3)^{\frac{1}{2}} = (a^{-7+3})^{\frac{1}{2}}$
$= (a^{-4})^{\frac{1}{2}}$
$= a^{-2}$
$= \dfrac{1}{a^2}$

d $\left(\dfrac{a^3b}{b^3}\right)^{-2} = \left(\dfrac{a^3}{b^2}\right)^{-2}$
$= \dfrac{a^{-6}}{b^{-4}}$
$= \dfrac{1}{a^6} \div \dfrac{1}{b^4}$
$= \dfrac{b^4}{a^6}$

Again the same answers can be obtained using the ability of some calculators to simplify expressions.

$5a^2y^3 \times 6a^4y^7$	
	$30 \cdot a^6 \cdot y^{10}$
$\dfrac{25a \times b^9}{15a^7 \times b^4}$	
	$\dfrac{5 \cdot b^5}{3 \cdot a^6}$

$(a^{-7} \times a^3)^{\frac{1}{2}}$	
	$\dfrac{1}{a^2}$
$\left(\dfrac{a^3b}{b^3}\right)^{-2}$	
	$\dfrac{b^4}{a^6}$

Exercise 1A

Evaluate each of the following without the use of a calculator.

1 $6^{20} \div 6^{18}$

2 $2^{19} \div 2^{16}$

3 $2^{12} \div 2^{8}$

4 3^{0}

5 5^{0}

6 $5^{0} + 2^{0}$

7 $(5+2)^{0}$

8 $\dfrac{2^3 \times 2^5}{2^7}$

9 $\dfrac{2^3 \times 2^5}{2^8}$

10 $\dfrac{2^{10}}{2^3 \times 2^4}$

11 1^{5}

12 $(-1)^{5}$

13 $(-2)^{5}$

14 $(-1)^{60}$

15 $(-1)^{61}$

16 $16^{\frac{1}{2}}$

17 $25^{\frac{1}{2}}$

18 $8^{\frac{1}{3}}$

19 $81^{\frac{1}{4}}$

20 $81^{\frac{1}{2}}$

21 3^{-2}

22 4^{-1}

23 2^{-3}

24 $2^{-1} + 4^{-1}$

25 $(2+4)^{-1}$

26 $9^{\frac{1}{2}} + 16^{\frac{1}{2}}$

27 $(9+16)^{\frac{1}{2}}$

28 $25^{-\frac{1}{2}}$

29 5^{0}

30 $(5^{0})^{2}$

31 $\left(5^{0}\right)^{\frac{1}{2}}$

32 $(-8)^{\frac{1}{3}}$

33 $25^{\frac{3}{2}}$

34 $\left(1\frac{7}{9}\right)^{\frac{1}{2}}$

35 $\left(2\frac{1}{4}\right)^{-\frac{1}{2}}$

36 $9^{\frac{3}{2}}$

37 $9^{-\frac{3}{2}}$

38 2^{-4}

39 $5^{0} + 2^{-1}$

40 $125^{\frac{2}{3}}$

41 $(-125)^{\frac{2}{3}}$

42 $64^{-\frac{2}{3}}$

43 $64^{-\frac{3}{2}}$

Express each of the following as a power of 2 (i.e. in the form 2^k).

44 $2^7 \times 2^9$

45 $2^6 \times 2^4 \times 2^3$

46 $2^9 \div 2^3$

47 $2^{11} \times 8$

48 $2^{11} \div 8$

49 1

50 $\dfrac{1}{2}$

51 $\dfrac{1}{2^3}$

52 $\dfrac{1}{8}$

ISBN 9780170390408

Express each of the following as a power of 3 (i.e. in the form 3^k).

53 27

54 81

55 1

56 $\frac{1}{3}$

57 $\sqrt{3}$

58 $\sqrt[4]{3}$

59 $\frac{1}{27}$

60 $\frac{1}{\sqrt{3}}$

61 $3 \times 3^2 \times 3^3$

Express each of the following as a power of 10 (i.e. in the form 10^k).

62 100

63 $\frac{1}{10}$

64 0.1

65 $\frac{1}{100}$

66 0.01

67 1

68 $(10^2)^3$

69 $(10^3)^2$

70 100^3

71 1000^3

72 $(0.1)^3$

73 $\sqrt{10}$

74 $(\sqrt{10})^6$

75 $\frac{1}{\sqrt{10}}$

76 $\sqrt{10^3}$

Determine the value of n in each of the following.

77 $2 \times 2 \times 2 \times 2 \times 2 = 2^n$

78 $3 \times 3 \times 3 \times 3 = 3^n$

79 $8 = 2^n$

80 $625 = 5^n$

81 $2^3 \times 2^4 = 2^n$

82 $3^8 \div 3^3 = 3^n$

83 $(3^2)^4 = 3^n$

84 $2^3 \times 2^4 \times 2 = 2^n$

85 $2^{15} \div 2^5 = 2^n$

86 $\frac{3^7 \times 3^4}{3^3} = 3^n$

87 $2^7 \times 4 = 2^n$

88 $\frac{2^4 \times 2^{11}}{2^7} = 2^n$

89 $(2^3 \times 2^2)^2 = 2^n$

90 $(2^3)^2 \times 2^2 = 2^n$

91 $2^3 \times (2^2)^2 = 2^n$

92 $6^3 = 2^n 3^n$

93 $\frac{9}{16} = \left(\frac{3}{4}\right)^n$

94 $3^4 \times 3^n \times 3 = 3^7$

95 $\frac{3^n}{3^4} = 3^{11}$

96 $(3^n)^5 = 3^{15}$

97 $(5^2)^n \times 5^3 = 5^{11}$

ISBN 9780170390408

Simplify each of the following without the assistance of a calculator and expressing your answers in terms of positive integers. (Then see if you can get the same answer on a calculator that has the ability to simplify expressions like these.)

98 $a^4 \times a^3$

99 $a^2 \div a^5$

100 $b^3 \div b^8$

101 $(b^3)^2$

102 $(b^{-2})^3$

103 $(a^{\frac{1}{2}})^4$

104 $(a^{-3})^2$

105 $a^{\frac{1}{2}} \times a^{\frac{3}{2}}$

106 $(b^2)^3 \times b^4$

107 $(b^2 \times b^4)^3$

108 $-4a^2 \times a^3$

109 $(-4a)^2 \times a^3$

110 $(a^2 \times a^{-3})^2$

111 $\left(\dfrac{a^3b^2}{b^3}\right)^3$

112 $2a^{-1} \times 3a^4$

113 $\dfrac{a^4 \times a}{a^8}$

114 $\dfrac{a^3b}{ab}$

115 $\dfrac{2a^4b}{a^3b}$

116 $4a^2 \times 2a^3$

117 $4a^2 \times (2a)^3$

118 $\dfrac{8a^3b^5}{2ab}$

119 $\dfrac{10ay^5}{a^6y^3}$

120 $\dfrac{12a^2b^7}{8a^6b^{10}}$

121 $\dfrac{6xy^2}{18x^2y}$

122 $\dfrac{(-3a^3b)^3}{3ab^2}$

123 $\dfrac{(a^2b^3)^2}{a^2b}$

124 $\dfrac{(3a^2)^2 \times b}{6b^3}$

125 $\dfrac{(a^4 \times a^{-12})^{\frac{1}{2}}}{a}$

126 $\left(\dfrac{x^4}{y^3}\right)^{-1}$

127 $\left(\dfrac{a^5b}{ab^3}\right)^{-2}$

128 $\dfrac{a^6 + a^3}{a^2}$

129 $\dfrac{a^7 + a^9}{a^2 \times a^3}$

130 $\dfrac{6a^4 + 9a^3}{3a^3}$

(Challenge):

131 $\dfrac{2^{n+2} + 12}{5 \times 2^n + 15}$

132 $\dfrac{2^{2n+3} - (2^n)^2}{2^n}$

133 $\dfrac{3^{n+1} + 9}{3^{n-1} + 1}$

ISBN 9780170390408

Solving equations involving indices

Solving exponential equations

Equations involving indices could have the unknown as the power or index, as in the following equations:

$2^x = 16,$ $\quad 3 \times 5^x = 375,$ $\quad 4^x - 5 = 11,$ $\quad 8^x = 4,$ $\quad 25^{3x-1} = 0.2$

or the unknown could be the base, as in the following:

$x^2 = 36,$ $\quad x^{0.5} = 9,$ $\quad x^{-1} = 9,$ $\quad 3 + x^{\frac{1}{2}} = 7,$ $\quad \frac{x}{\sqrt{x}} = 7$

Questions 77 to **97** in the previous exercise all involved equations in which the unknown, in that case n, featured as an index. The next example shows a few more of this type.

EXAMPLE 5 (UNKNOWN AS THE INDEX)

Solve the following equations

a $2^x = 16$ **b** $3 \times 5^x = 375$ **c** $4^x - 5 = 11$ **d** $8^x = 4$ **e** $25^{3x-1} = 0.2$

Solution

a Given: $2^x = 16$

An awareness of the powers of 2, i.e. $2^2 = 4$, $2^3 = 8$, $2^4 = 16$, … enables the equation to be solved mentally: $x = 4$

b Given: $3 \times 5^x = 375$

Divide each side by 3: $5^x = 125$

From an awareness of the powers of 5: $x = 3$

c Given: $4^x - 5 = 11$

Add 5 to both sides to isolate 4^x: $4^x = 16$

From an awareness of the powers of 4: $x = 2$

d Given: $8^x = 4$

You may again be able to determine the answer intuitively but, if not, notice that we can express each side of the equation as powers of the same base, in this case 2: $(2^3)^x = 2^2$

i.e. $2^{3x} = 2^2$

Hence $3x = 2$

∴ $x = \frac{2}{3}$

e Given: $25^{3x-1} = 0.2$

Noticing that 25 and 0.2 can both be expressed as powers of 5: $(5^2)^{3x-1} = 5^{-1}$

i.e. $5^{6x-2} = 5^{-1}$

Hence $6x - 2 = -1$

$6x = 1$

∴ $x = \frac{1}{6}$

ISBN 9780170390408

Alternatively, the previous equations could be solved using the ability of some calculators to solve equations.

```
solve(2^x = 16, x)
                        {x = 4}
solve(3 × 5^x = 375, x)
                        {x = 3}
solve(4^x – 5 = 11, x)
                        {x = 2}
solve(8^x = 4, x)
                        {x = 2/3}
solve(25^(3x – 1) = 0.2, x)
                        {x = 1/6}
```

If the unknown is the base, the technique is to steadily reduce the power of the unknown by performing suitably chosen operations to both sides of the equation, as shown in the next example.

It must be remembered though that when reducing $x^n = c$ to $x = \sqrt[n]{c}$, then if n is even we must say that $x = \pm\sqrt[n]{c}$.

For example, if $x^2 = 64$ then $x = \pm\sqrt{64} = \pm 8$.

But if $x^3 = 64$ then $x = \sqrt[3]{64} = 4$.

EXAMPLE 6 (UNKNOWN AS THE BASE)

Solve the following equations.

a $x^2 = 36$ **b** $x^{0.5} = 9$ **c** $x^{-1} = 9$

d $3 + x^{\frac{1}{2}} = 7$ **e** $\dfrac{x}{\sqrt{x}} = 7$

Solution

a Given: $x^2 = 36$

Take the square root of each side: $x = \pm\sqrt{36}$

$= \pm 6$

b Given: $x^{0.5} = 9$

i.e. $\sqrt{x} = 9$

Square each side: $(x^{0.5})^2 = 9^2$

i.e. $x = 81$

Shutterstock.com/Feng Yu

ISBN 9780170390408

c Given: $x^{-1} = 9$

i.e. $\frac{1}{x} = 9$

Multiply each side by x: $1 = 9x$

Divide by 9 to isolate x: $\frac{1}{9} = x$

i.e. $x = \frac{1}{9}$

Alternatively: Raise each side of the initial equation to the power −1 to give $x = 9^{-1}$

i.e., as before, $x = \frac{1}{9}$

Again, these same equations could be solved using the equation solving ability of some calculators.

d Given: $3 + x^{\frac{1}{2}} = 7$

Subtract 3 from each side: $x^{\frac{1}{2}} = 4$

Squaring each side gives: $x = 16$

e Given: $\frac{x}{\sqrt{x}} = 7$

i.e. $x^{1-0.5} = 7$

$x^{0.5} = 7$

Squaring each side gives: $x = 49$

Exercise 1B

Solve each of the following equations without the help of a calculator.

Unknown in the index

1 $2^x = 8$ **2** $2^x = 32$ **3** $2^x = 128$ **4** $2^x = \frac{1}{8}$

5 $2^x = \frac{1}{32}$ **6** $2^x = \frac{1}{128}$ **7** $2^a = \sqrt{2}$ **8** $2^a = \frac{1}{\sqrt{2}}$

9 $2^y = \frac{1}{4}$ **10** $5^c = 125$ **11** $10^d = 1000$ **12** $4^x - 3 = 61$

13 $3^x - 2 = 25$ **14** $2^y - 6 = 58$ **15** $2 \times 5^x = 50$ **16** $3^{2x} = 9$

17 $5^{4x} = 125$ **18** $5^{4+x} = 125$ **19** $3 \times 2^x = 24$ **20** $\frac{10^x}{5} = 20$

ISBN 9780170390408

21 $\frac{2^x}{4} = 8$

22 $16^k = 8$

23 $16^p = \frac{1}{2}$

24 $9^x = 27$

25 $2^{3x+1} = 16$

26 $2^{15-2h} = 8$

27 $4^{x-1} = 0.5$

28 $3^{x+1} = \frac{27}{3^x}$

29 $16^{x+2} = 128$

30 $5^{2n-1} = 125$

Unknown in the base

31 $a^2 = 16$

32 $p^2 = 100$

33 $x^3 = 8$

34 $x^3 = 64$

35 $x^{\frac{1}{2}} = 4$

36 $x^{\frac{1}{3}} = 4$

37 $h^{-1} = 4$

38 $y^{-1} = 2$

39 $p^{-1} = \frac{1}{3}$

40 $x^{0.5} = 100$

41 $3x^2 = 75$

42 $9x^2 = 4$

43 $x^4 + 7 = 88$

44 $3 + x^{\frac{1}{3}} = 13$

45 $p^2 - 3 = 13$

46 $\frac{x^5}{x^2} = 64$

47 $\frac{x}{\sqrt{x}} = 9$

48 $\frac{x^2}{x^3} = 16$

49 $(w-2)^3 = 8$

50 $(2x-1)^3 = 27$

51 $(h+1)^{\frac{1}{2}} = 5$

52 $(x-3)^2 = 16$

53 $2w^{\frac{1}{2}} = 3$

54 $2z^{\frac{1}{3}} = 3$

55 **a** Consider the equation $x^5 = 16x^3$.

Dividing both sides by x^3 gives $\frac{x^5}{x^3} = \frac{16x^3}{x^3}$,

i.e., $x^2 = 16$,

hence $x = \pm 4$.

However, as well as these two solutions of $x = 4$ and $x = -4$, there is another value for x that satisfies the original equation. What is this third solution to the original equation?

b Instead of dividing each side of the original equation by x^3 to solve it, subtract $16x^3$ from each side and then factorise and solve.

c Solve each of the following equations:

i $x^2 = 4x$ **ii** $x^3 = 25x$ **iii** $x^3 = 25x^2$ **iv** $x^3 + 16x = 0$

ISBN 9780170390408

What if we cannot solve the equation mentally or algebraically?

Asked to solve the equation $8^x = 4$

Our familiarity with powers of 2 allows us to write this as $(2^3)^x = 2^2$

i.e. $2^{3x} = 2^2$

Hence $3x = 2$

and so $x = \frac{2}{3}$

Similarly, given the equation $25^{3x-1} = 0.2$

Our familiarity with powers of 5 allows us to write this as $5^{6x-2} = 5^{-1}$

Giving $x = \frac{1}{6}$

However, suppose we were asked to solve $2^x = 11$ or perhaps $5^{2x-3} = 48$.

In such cases we could use one of the following:

- The solve facility on some calculators
- A graphical approach
- Trial and adjustment
- Logarithms, a concept that will be introduced in a later unit of *Mathematical Methods*.

Using the solve facility on some calculators

```
solve(2^x = 11, x)
        {x = 3.459431619}
solve(5^(2x - 3) = 48, x)
        {x = 2.702656213}
```

A graphical approach

To solve $2^x = 11$

draw $y = 2^x$

and see where $y = 11$

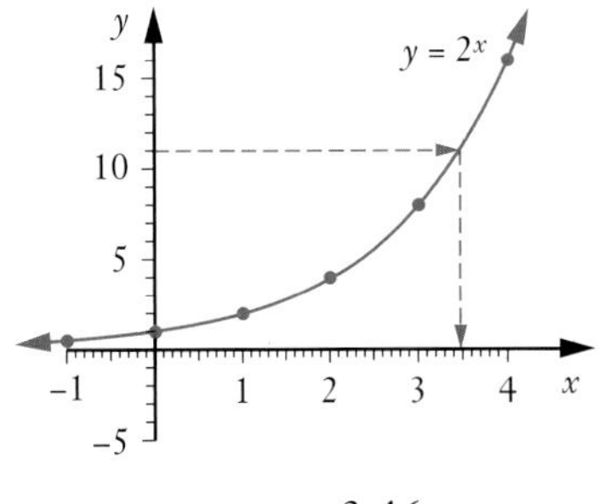

$x \approx 3.46$

To solve $5^{2x-3} = 48$

draw $y = 5^{2x-3}$

and see where $y = 48$

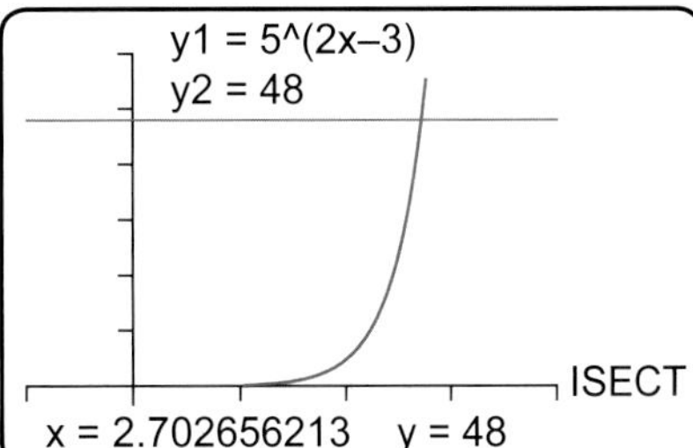

$x = 2.703$ (to 3 decimal places)

ISBN 9780170390408

To solve $2^x = 11$

Note that $2^3 = 8$

and $2^4 = 16$

Try $x = 3.5$ $\quad 2^{3.5} \approx 11.31$ (by calculator)

Try $x = 3.4$ $\quad 2^{3.4} \approx 10.56$ (by calculator)

Try $x = 3.45$ $\quad 2^{3.45} \approx 10.93$ (by calculator)

Thus, for $2^x = 11$, x must lie between 3.45 and 3.5.

Therefore, correct to one decimal place, $x = 3.5$.

To solve $5^{2x-3} = 48$

Note that $5^2 = 25$

and $5^3 = 125$

Try $x = 2.5$ $\quad 5^{2.5} \approx 55.9$

Try $x = 2.4$ $\quad 5^{2.4} \approx 47.59$

Try $x = 2.45$ $\quad 5^{2.45} \approx 51.58$

Thus, correct to 1 decimal place,

$$2x - 3 = 2.4$$
$$2x = 5.4$$
$$x = 2.7$$

Exercise 1C

Trial and adjustment

Solve each of the following equations using 'trial and adjustment', giving answers correct to one decimal place.

1 $2^x = 23$

2 $3^x = 33$

3 $5^x = 50$

4 $7^x + 5 = 245$

5 $2^{x-1} = 51$

6 $3^{x+2} = 100$

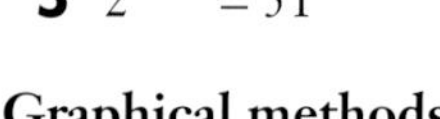

Shutterstock.com/julie deshales

Graphical methods

Solve each of the following equations using the ability of some calculators to graph functions. (Give answers correct to two decimal places.)

7 $3^x = 15$

8 $5^x = 61$

9 $4^x = 100$

10 $2^x + x = 83$

11 $2^x + 3^x = 84$

12 $5^x - 3^x = 250 - x$

The solve facility on some calculators

Solve each of the following using the solve facility on some calculators. (Give answers correct to two decimal places.)

13 $2^x = 345$

14 $2^x = 0.35$

15 $5^x = 1100$

16 $2^{2x-1} = 51$

17 $7 + 5^x = 89$

18 $2^x - x = 7$

ISBN 9780170390408

Miscellaneous exercise one

This miscellaneous exercise may include questions involving the work of this chapter and the ideas mentioned in the Preliminary work section at the beginning of the book.

1 Copy and complete the following table of values for the general cubic $y = ax^3 + bx^2 + cx + d$.

x	0	1	2	3	4
y		$a+b+c+d$			
1st difference	$a+b+c$				
2nd difference					
3rd difference					

Hence determine the equation for the function having the following table of values.

x	0	1	2	3	4	5	6
y	7	7	5	−5	−29	−73	−143

2 Rewrite each of the following sentences with the number written in standard form, or scientific notation, written in 'long hand'.

- **a** Australia has an area of approximately 7.682×10^6 km^2.
- **b** Light travels at a speed of 3×10^8 m/s.
- **c** A golf ball has a mass of approximately 4.5×10^{-2} kg.
- **d** Earth is approximately 1.5×10^8 km from the sun.
- **e** Gamma waves have a wavelength of less than 10^{-11} metres.
- **f** The Earth orbits the sun at a speed of approximately 1.07×10^5 km/h.
- **g** In 1961 the first man in space, Yuri Gagarin, flew his spacecraft at a speed of 2.74×10^4 km/h, i.e. approximately 7.6×10^3 m/h.

Getty Images/Science Photo Library

3 Rewrite each of the following sentences with the 'long hand' number written in standard form, or scientific notation, i.e. in the form $A \times 10^n$ where A is a number between 1 and 10 and n is an integer.

- **a** At the beginning of this century China had a population of approximately 1 270 000 000 and India had a population of approximately 1 030 000 000.
- **b** The egg cell, or ovum, with a radius of approximately 0.000 05 metres, i.e. 0.05 mm, is the largest single human cell.
- **c** It is thought that approximately 1 100 000 people die each year of malaria.
- **d** Some adult wasps of a particular species could weigh just 0.005 grams.
- **e** Concorde, the first supersonic passenger airliner, had a cruising speed of 2160 km/h.

ISBN 9780170390408

4 Round each of the following to the stated number of significant figures.

a 12 432 000 to 2 significant figures

b 46 790 to 3 significant figures

c 304 702 125 to 3 significant figures

d 0.012 04 to 1 significant figure

e 0.205 701 to 3 significant figures

f 0.005 607 to 1 significant figure

Shutterstock.com/Senohrabek

5 Express each of the following in the form 5^k.

a 25 **b** 125 **c** $\sqrt{5}$ **d** $\frac{1}{5}$

e $\frac{1}{25}$ **f** $\frac{1}{\sqrt{5}}$ **g** $5^3 \times 5^4$ **h** $5 \times 5^4 \times 25$

i $5^8 \div 5^2$ **j** $(5^3)^4$ **k** $(25)^3$ **l** 1

m $\frac{5^7 \times 5^1}{5^3}$ **n** $\frac{5^9}{5^3 \times 5^4}$ **o** $\frac{5^7}{15+10}$

6 Solve each of the following equations.

a $2^x = 32$ **b** $5^x = 625$ **c** $3^x = \frac{1}{9}$ **d** $(2^2)^x = 32$

e $8^x = 32$ **f** $64^x = 4$ **g** $125^x = \frac{1}{5}$ **h** $5^x = \frac{1}{125}$

i $2^{5x} = \frac{1}{4}$ **j** $25^x = \sqrt{5}$ **k** $49^x = \frac{1}{343}$ **l** $3^x = \frac{3^{10}}{9}$

7 Solve each of the following equations.

a $y^{-2} = 9$ **b** $p^{\frac{1}{2}} = \frac{2}{3}$ **c** $x^{-1} = \frac{2}{3}$ **d** $x^{\frac{1}{3}} = 2$

e $x^2 = 25$ **f** $t^{-2} = 25$ **g** $(2t)^{\frac{1}{3}} = 3$ **h** $3x^{\frac{1}{2}} + 2x^{\frac{1}{2}} = 15$

i $x^3 = x^2$ **j** $x^3 = x$

8 Solve the following equations:

a $(2^x)^2 - 10 \times 2^x + 16 = 0$ **b** $3^{2x} - 10 \times 3^x + 3^2 = 0$

c $2^{2x+1} - 3 \times 2^x + 1 = 0$ **d** $2^{2x-1} - 5 \times 2^x + 8 = 0$

ISBN 9780170390408

Exponential functions

- Exponential relationships
- Growth and decay
- Miscellaneous exercise two

Situation One

The number of cells in a particular organism increases by cell division. In this process one cell splits into two cells which in turn each split into two cells and so on.

Let us suppose that there is initially 1 cell and that this cell division occurs approximately every week, i.e. the number of cells present doubles every week.

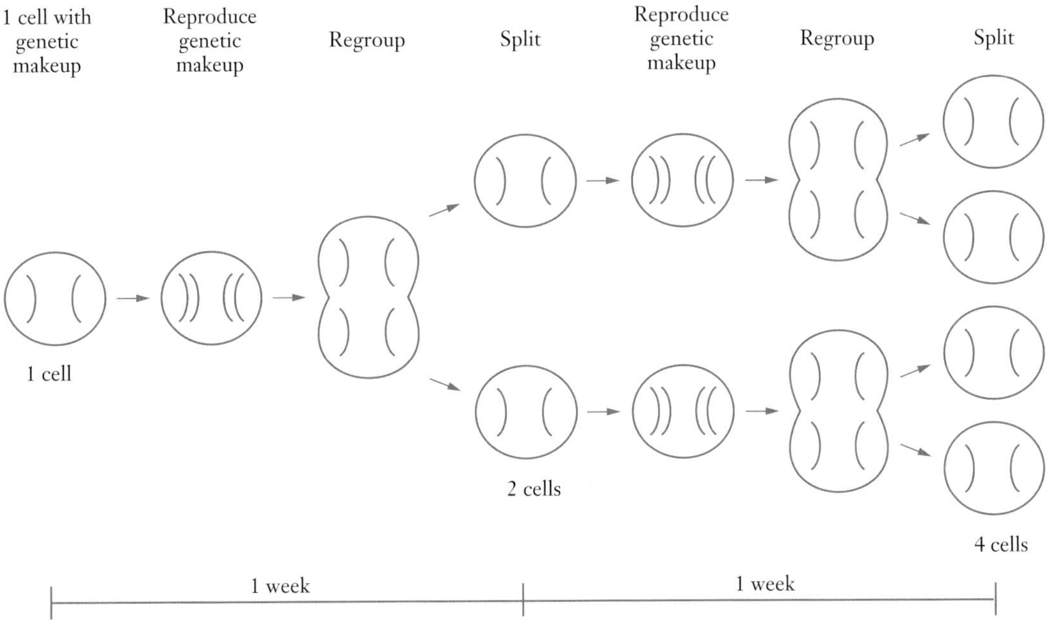

Copy and complete the following table.

Number of weeks, *t*	0	1	2	3	4	5	6	7	8	9	10
Number of cells, *C*	1	2	4								

a How long does it take for there to be 100 cells present?

b How long does it take for there to be 400 cells present?

c Determine the rule for C in terms of t.

d Let us suppose that in a very simplified model of the growth of a human baby, we assume that the initial single cell divides into two cells after one week. After another week, these two each divide into two to give four cells altogether. If this cell division continues each week, how many cells are there after 40 weeks?

Now read Situations two to five that follow. They do not ask you to do any calculations. Simply read them and make sure that you agree with, and understand, what is said.

Situation Two

Consider an investment of \$500 earning interest of 10% compounded annually.

The value of this investment for the first eight years is shown tabulated and graphed below

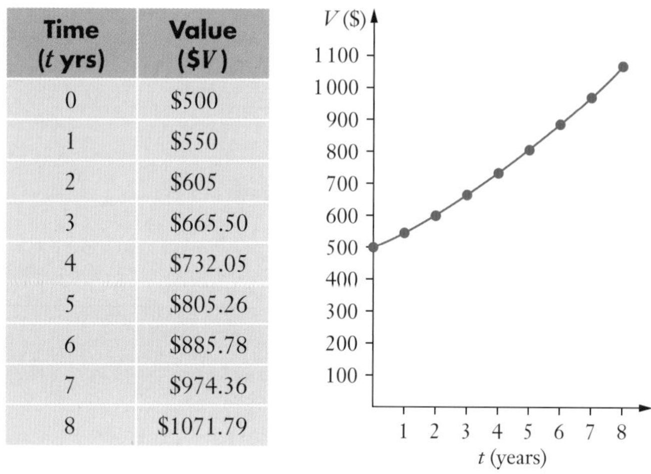

Time (t yrs)	Value (\$$V$)
0	\$500
1	\$550
2	\$605
3	\$665.50
4	\$732.05
5	\$805.26
6	\$885.78
7	\$974.36
8	\$1071.79

The figures in the V column commence with 500 and then each one thereafter is the previous one multiplied by 1.1.

Situation Three

Consider a culture of bacteria with an initial population of 100 cells with the number doubling every hour.

This population for the first seven hours is shown tabulated and graphed below.

Time (t hrs)	Population (P)
0	100
1	200
2	400
3	800
4	1 600
5	3 200
6	6 400
7	12 800

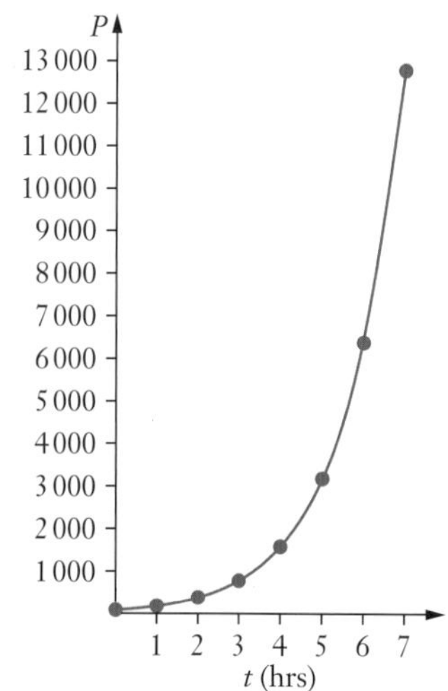

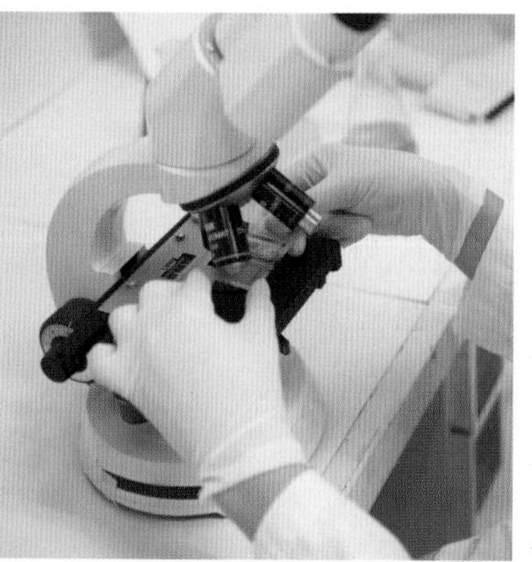

Shutterstock.com/Apples Eyes Studio

The figures in the P column commence with 100 and then each one thereafter is the previous one multiplied by 2.

ISBN 9780170390408

Situation Four

Consider a car that has an initial value of $40 000 and by the end of each year it has lost 12% of what its value was at the beginning of that year. The table of values for the first seven years, the rule for the situation, and the graph, are shown below.

Shutterstock.com/Andrey Lobachev

Time (t yrs)	Value (\$$V$)
0	$40000
1	$35200
2	$30976
3	$27259
4	$23988
5	$21109
6	$18576
7	$16347

Rule

Value after t years is given by:

$$V = 40\,000 \times 0.88^t$$

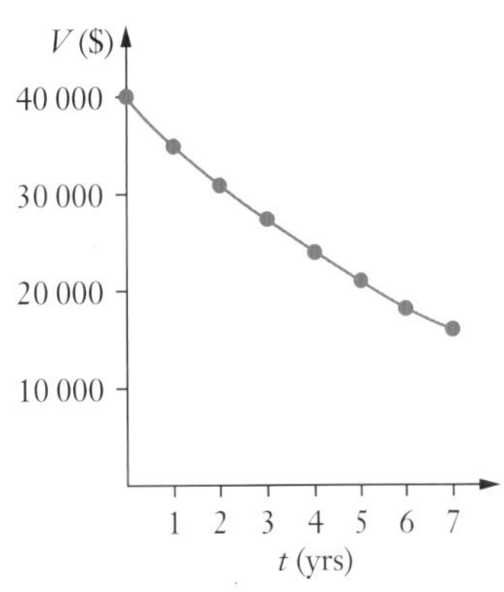

Situation Five

Consider a radioactive element decaying at a rate that sees 40% of the element decay to a more stable form each hour. Thus 500 g of the element becomes 300 g (= 60% of 500 g) one hour later, 180 g (= 60% of 300 g) one hour after that, and so on. The table of values for the first seven hours, the rule for the situation, and the graph, are shown below.

Shutterstock.com/Pi-Lens

Time (t yrs)	Amount (A g)
0	500
1	300
2	180
3	108
4	65
5	39
6	23
7	14

Rule

Amount present after t hours is given by:

$$A = 500 \times 0.6^t$$

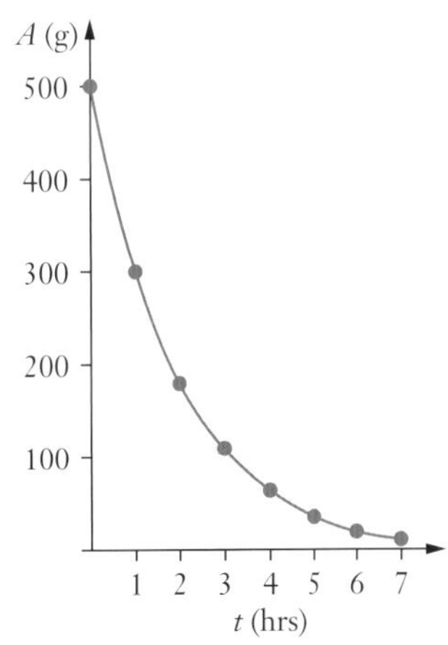

Exponential functions

Translating exponential graphs

Graphing exponentials

Exponential relationships

The situations on the previous pages involved a quantity being repeatedly multiplied by a number. The situations were all examples of **exponential relationships**.

In such relationships, because we repeatedly multiply by a number, the **ratio** of successive entries will be constant (rather than the first *difference pattern* being constant, as in linear, or the second *difference pattern* being constant, as in quadratic).

For example, suppose we make a sequence of numbers for which the first is 3 and we repeatedly multiply by 2:

	3	6	12	24	48	96	192	
1st difference	3	6	12	24	48	96		← Not const. Hence not linear.
Ratio	$\frac{6}{3}=2$	$\frac{12}{6}=2$	$\frac{24}{12}=2$	$\frac{48}{24}=2$	$\frac{96}{48}=2$	$\frac{192}{96}=2$		← Constant ratio. Exponential.

Each input value gives one, and only one, output value. The relationship is therefore a function. (Notice that the graphs on the previous pages pass the vertical line test.)

Exponential functions are characterised by **rules** of the form $y = y_0\, a^x$, $a > 0$.

- y_0 is the value of y when $x = 0$,
- a is the constant multiplying factor.

For example, the rule $y = 3.5 \times 4^x$ generates the following table of values.

x	0	1	2	3	4	5	6
y	3.5	14	56	224	896	3584	14336
1st ratio	$\frac{14}{3.5}=4$	$\frac{56}{14}=4$	$\frac{224}{56}=4$	$\frac{896}{224}=4$	$\frac{3584}{896}=4$	$\frac{14336}{3584}=4$	

The **graphs** of exponential functions have the characteristic shape shown on the right by the graph of

$$y = 2^x$$

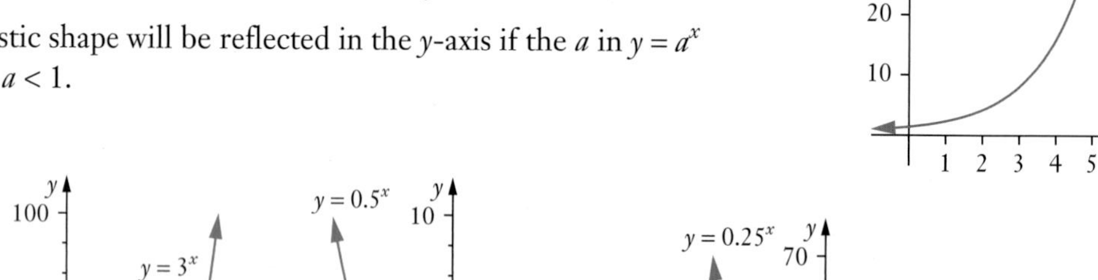

This characteristic shape will be reflected in the y-axis if the a in $y = a^x$ is such that $0 < a < 1$.

For example:

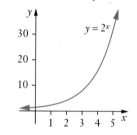

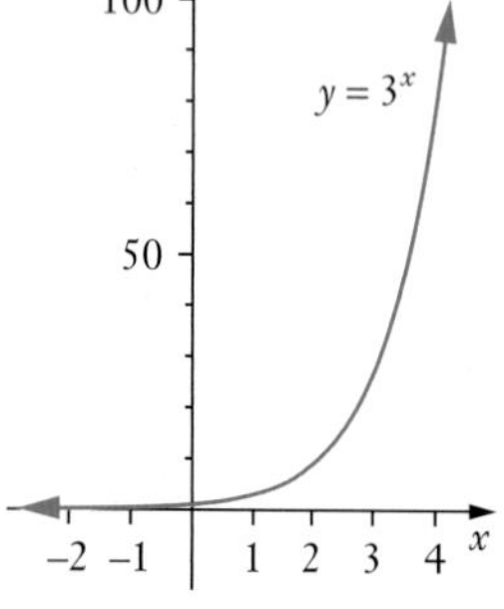

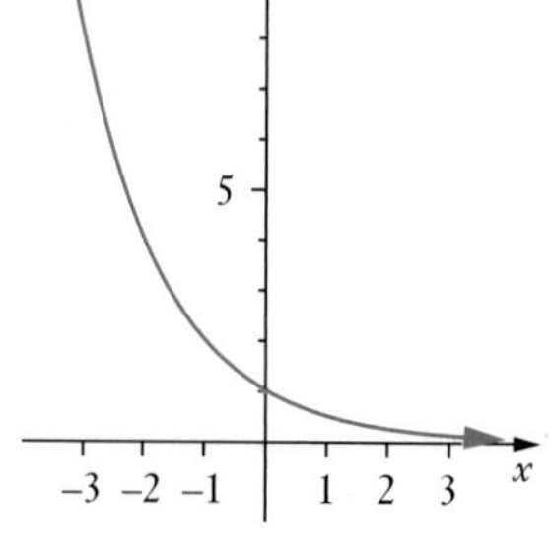

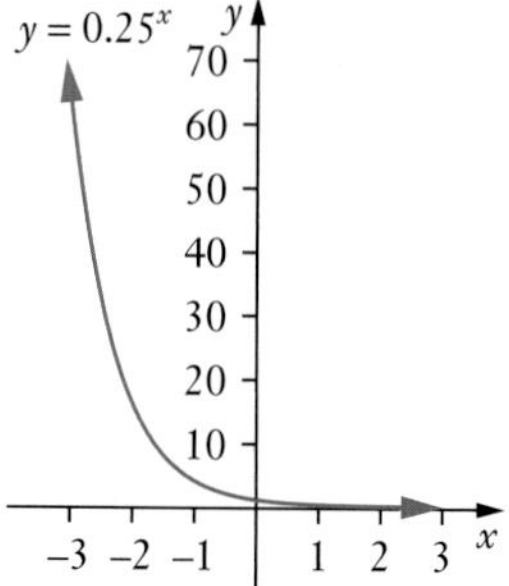

ISBN 9780170390408

Exercise 2A

For questions 1 to 6, copy and complete the following tables for the exponential rule stated.

1 Rule: $y = 3^x$

x	0	1	2	3	4	5
y						

2 Rule: $y = 7^x$

x	0	1	2	3	4	5
y						

3 Rule: $y = 1.5 \times 2^x$

x	0	1	2	3	4	5
y						

4 Rule: $y = 1.75 \times 8^x$

x	0	1	2	3	4	5
y						

5 Rule: $y = 2^{x+1}$

x	0	1	2	3	4	5
y						

6 Rule: $y = 2.5 \times 4^{x+1}$

x	1	2	3	4	5	6
y						

For each of the tables shown in questions 7 to 18 below

a determine whether the function involved is linear, quadratic, cubic, exponential, reciprocal, or none of these.

b For those that are one of the five types mentioned, determine the equation of the function.

7

x	0	1	2	3	4
y	1	2	5	10	17

8

x	0	1	2	3	4
y	1	4	16	64	256

9

x	0	1	2	3	4
y	3	5	7	9	11

10

x	0	1	2	3	4
y	0	2	8	18	32

11

x	0	1	2	3	4
y	1.5	12	96	768	6144

12

x	0	1	2	3	4
y	1	5	25	125	625

13

x	0	1	2	3	4
y	0	2	6	12	20

14

x	0	1	2	3	4
y	1	6	36	216	1296

15

x	0	1	2	3	4
y	3	6	12	24	48

16

x	1	2	3	4	5
y	60	30	20	15	12

17

x	0	1	2	3	4
y	1	2	9	28	65

18

x	0	1	2	3	4
y	20	17	14	11	8

ISBN 9780170390408

19 a Display the graphs of the following exponential functions on a graphic calculator using an x-axis from 0 to 5 and a y-axis from 0 to 40.

$$y = 1.25^x$$
$$y = 1.5^x$$
$$y = 1.75^x$$
$$y = 2^x$$
$$y = 3^x$$

State the coordinates of the point that all of these functions pass through.

b Write a few sentences, including sketches if you wish, to describe the characteristic shape of the graphs of functions of the form $y = a^x$ $(a > 1)$ and describe the effect that changing the value of a has on the graph.

20 a Display the graphs of the following exponential functions on a graphic calculator using an x-axis from –3 to 4 and a y-axis from –1 to 10.

$$y = 2^x$$
$$y = 2(2)^x$$
$$y = 3(2)^x$$
$$y = 4(2)^x$$

b Write a few sentences, including sketches if you wish, to describe the effect that increasing the value of a has on the graph of $y = a(2)^x$ for $a \geq 1$.

21 Investigate the effect that changing the value of k has on the graph of

$$y = a^x - k.$$

22 Investigate the effect that changing the value of k has on the graph of

$$y = a^{x-k}.$$

23 Each graph shown below is of the form $y = a^x$, for integer a.

Find the equation of each.

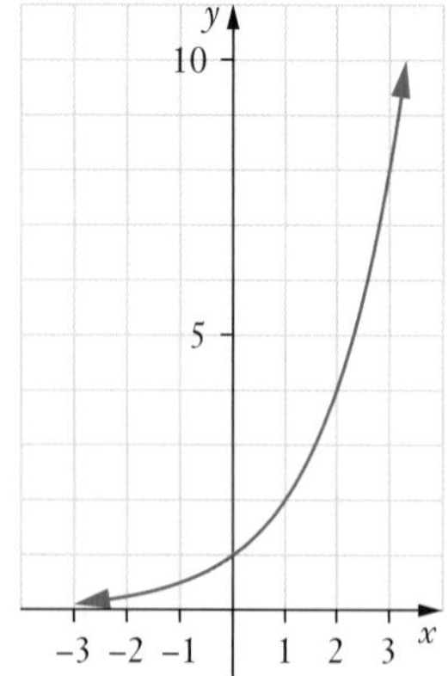

b

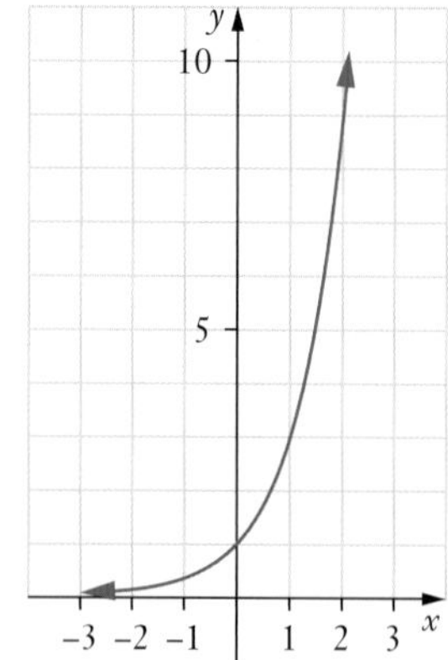

ISBN 9780170390408

24 The population of a country is growing such that in t years' time, the population will be P million, where $P \approx 25(1.04)^t$. Display the graph of $P = 25(1.04)^t$, for $0 \leq t \leq 50$, on a graphic calculator.

Using your graph, or by other methods, predict in how many years the population will be

a 40 000 000 **b** 75 000 000 **c** 120 000 000.

25 Determine the equations of each of the exponential functions shown graphed below given that each is a *translation* of either $y = 2^x$ or $y = 3^x$.

a

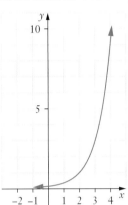

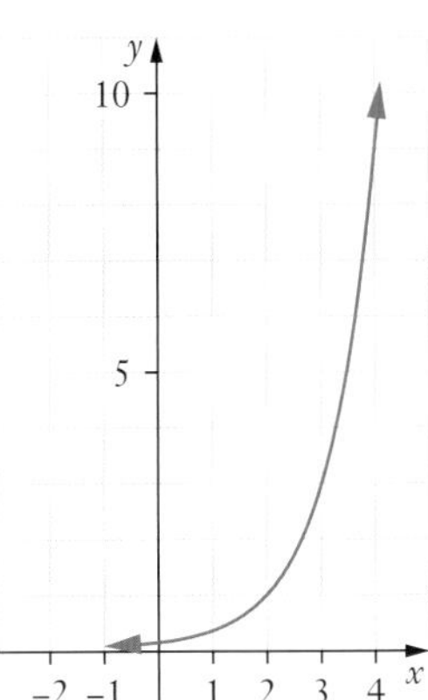

b

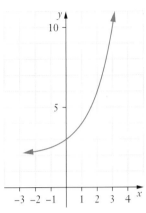

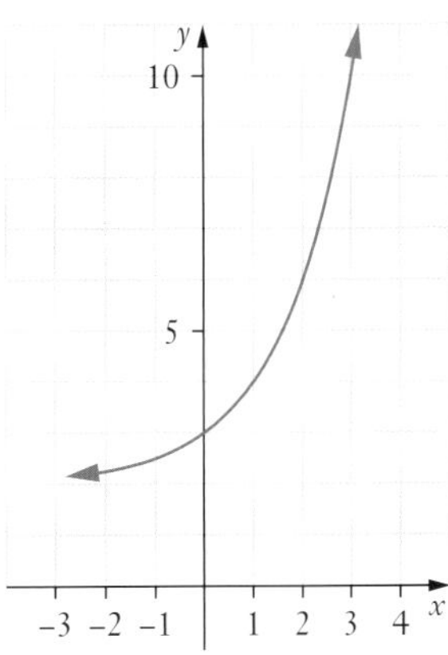

c

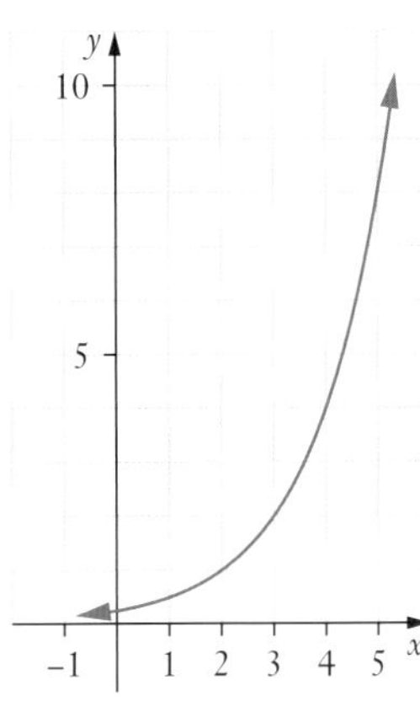

d

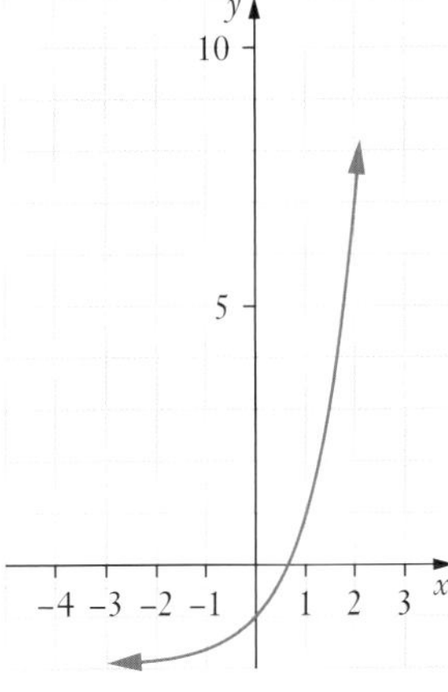

e

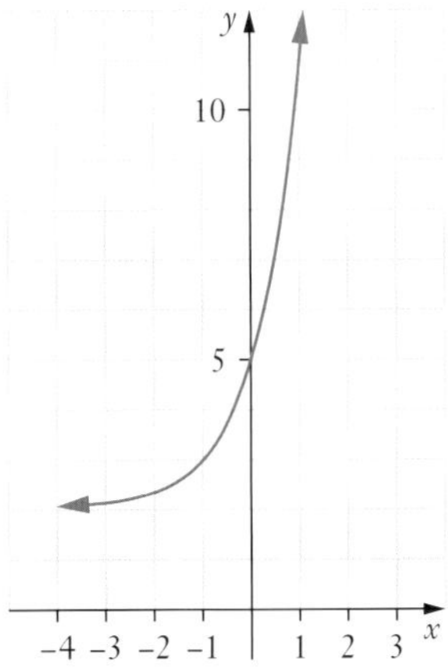

f

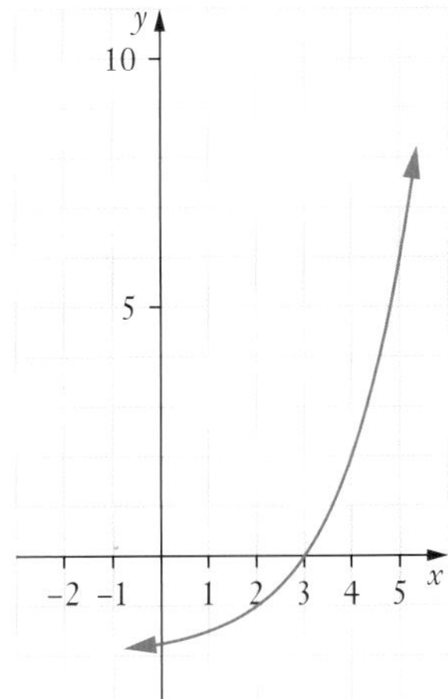

ISBN 9780170390408

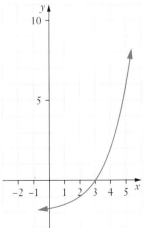

Using exponential models

Growth and decay

Thinking again about Situations one to five at the beginning of this chapter:

Situation one, the cell splitting, involved a *growth* situation. $C = 2^t$
Situation two, the investment, was a *growth* situation. $V = 500 \times 1.1^t$
Situation three, the culture of bacteria, was a *growth* situation. $P = 100 \times 2^t$

Situation four, the value of a car, was a *decay* situation. $V = 40\,000 \times 0.88^t$
Situation five, the radioactive element, was a *decay* situation. $A = 500 \times 0.6^t$

Exponential *growth* is characterised by an equation of the form $y = y_0 a^x, \quad a > 1.$

Exponential *decay* is characterised by an equation of the form $y = y_0 a^x, \quad 0 < a < 1.$

EXAMPLE 1

The population of a country is growing exponentially. The populations in 2010, 2011, 2012 and 2013 were as follows.

Year	2010	2011	2012	2013
Population	20 million	20.4 million	20.81 million	21.22 million

Predict the population for this country in the year 2030.

Solution

First find the annual growth rate: $\frac{20.4}{20} = 1.02$

$$\frac{20.81}{20.4} \approx 1.0201$$

$$\frac{21.22}{20.81} \approx 1.0197$$

Each year the population is multiplied by roughly 1.02 (i.e., a 2% increase per year).

Thus, by the year 2030 the population will be roughly

$$\begin{aligned} 20 \text{ million} \times 1.02^{(2030-2010)} &= 20 \text{ million} \times 1.02^{20} \\ &\approx 29.7 \text{ million} \end{aligned}$$

The population for this country will be approximately 29.7 million by the year 2030 (assuming the growth rate shown in the given years continues).

ISBN 9780170390408

EXAMPLE 2

The population of a particular endangered species of animal is declining exponentially. The populations in 2000, 2005, 2010 and 2013 were thought to be as follows.

Year	2000	2005	2010	2013
Population	5760	4460	3450	2960

a By what percentage is the population declining each year?

b Predict the population for this animal in the year 2025.

Solution

a Consider the period 2000 to 2005.

If each year the population is multiplied by r, then $5760r^5 = 4460$

$\therefore$ $r^5 = \dfrac{4460}{5760}$

Thus $r = \sqrt[5]{\dfrac{4460}{5760}}$

≈ 0.95

Consider the period 2005 to 2010.

If each year the population is multiplied by r, then $4460r^5 = 3450$

$\therefore$ $r^5 = \dfrac{3450}{4460}$

Thus $r = \sqrt[5]{\dfrac{3450}{4460}}$

≈ 0.95

Thus, each year the population is multiplied by 0.95 (i.e. a 5% decrease per year).

Check: For 2010 to 2013, $3450 \times 0.95^3 \approx 2960$, as required.

The population is falling by 5% each year.

b By the year 2025, the population will be roughly

$$5760 \times 0.95^{(2025 - 2000)}$$
$$= 5760 \times 0.95^{25}$$
$$\approx 1600$$

By the year 2025, the population will be roughly 1600 (assuming the rate of decline for the given years continues).

ISBN 9780170390408

EXAMPLE 3

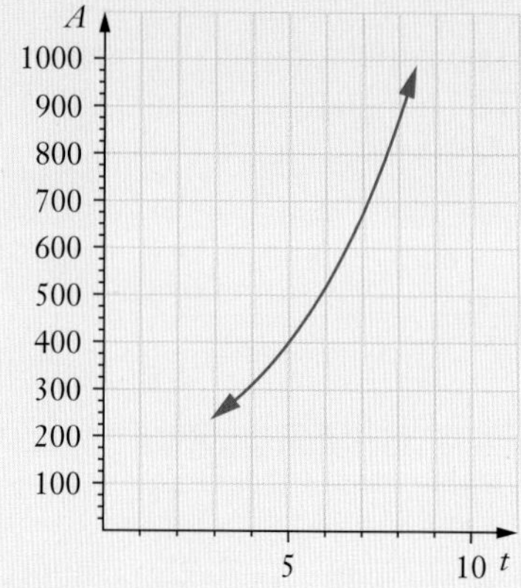

The graph on the right shows an exponential growth situation with the variables A and t related according to a rule of the form

$$A = ka^t \text{ for } a > 1.$$

Determine

a the value of A when $t = 5$,

b the value of A when $t = 8$,

c the constants a and k,

d the value of A when $t = 0$,

e the value of t (correct to one decimal place) for which $A = 4000$, assuming that the growth rate suggested by the graph continues.

Solution

a From the graph, when $t = 5$, $A \approx 400$.

b From the graph, when $t = 8$, $A \approx 875$.

c Exponential growth is involved. Thus, from $t = 5$ to $t = 8$ we have multiplied by a three times.

Thus
$$400a^3 = 875$$
$$a = \sqrt[3]{2.1875}$$
$$\approx 1.298$$

The relationship is of the form $A = k(1.298)^t$

But when $t = 5, A \approx 400$.
$$\therefore 400 = k(1.298)^5$$
$$k \approx 109$$

Thus $a \approx 1.298$ and $k \approx 109$.

d The relationship is of the form $A \approx 109(1.298)^t$

Thus when $t = 0$
$$A \approx 109(1.298)^0$$
$$= 109$$

When $t = 0, A \approx 109$.

e The relationship is of the form $A \approx 109(1.298)^t$

Thus, when $A = 4000$
$$4000 \approx 109(1.298)^t$$
$$(1.298)^t \approx 36.70$$

Solving by calculator or by trial and adjustment $t = 13.8$, correct to one decimal place.

Thus $A \approx 4000$ when $t = 13.8$.

Note: With $A = ka^t$ then when $t = 0, A = k$. Thus had the graph in the above example shown where the curve cut the vertical axis, this point would have given us the value of k directly.

ISBN 9780170390408

Exercise 2B

1 Show that the following figures support the claim that the annual percentage growth rate is approximately 8%.

Year	1995	2000	2010
Population	18 000 000	26 000 000	56 000 000

2 Show that the following figures support the claim that the annual percentage decay rate is approximately 4.5%.

Year	2000	2007	2011
Population	12 400	9000	7500

3 The population of a country is growing exponentially. The populations in 2010, 2011, 2012 and 2013 were as follows.

Year	2010	2011	2012	2013
Population	45 million	45.8 million	46.6 million	47.5 million

Predict the population for this country in the year 2027.

4 The population of a particular species of animal is declining exponentially. The numbers of these animals thought to be in existence in the wild in 2010, 2011, 2012 and 2013 were as follows.

Year	2010	2011	2012	2013
Number	18 000	16 500	15 200	14 000

If the above figures are correct, and nothing is done to alter the rate of decline, how many of these animals will exist in the wild in the year 2023?

5 An analysis of the membership of a particular sports club since it was first formed in 1989 indicated that the membership each year could have been quite accurately predicted using the exponential model:

$$\text{Membership in the year } N \approx Ak^{(N-1989)}.$$

The number of members initially and on the tenth, the twentieth and the twenty fifth anniversaries of the founding of the club were as follows.

Year	1989	1999	2009	2014
Members	80	170	375	550

a Find the values of A and of k (state k to 2 decimal places).

b By what percentage is the membership growing each year?

c Predict the number of members for the year 2024 (nearest hundred).

ISBN 9780170390408

6 During a drought a particular river bed dries up and the colony of frogs living in the vicinity experiences an exponential population decline.

The estimated number of frogs in the colony, t days after drought conditions were officially declared, was as follows.

t	5	6	7	8
Population	530	450	385	325

According to these estimated figures what was the population of the frog colony initially (i.e., at $t = 0$)?

7 The graph shows an exponential decay situation with the variables P and t related according to a rule of the form:

$$P = ka^t \quad \text{for} \quad 0 < a < 1.$$

Determine

a the value of P when $t = 3$,

b the value of P when $t = 8$,

c the constants a (2 decimal places) and k (nearest 5),

d the value of P when $t = 0$ (nearest 5),

e the value of t (nearest integer) for which $P = 10$.

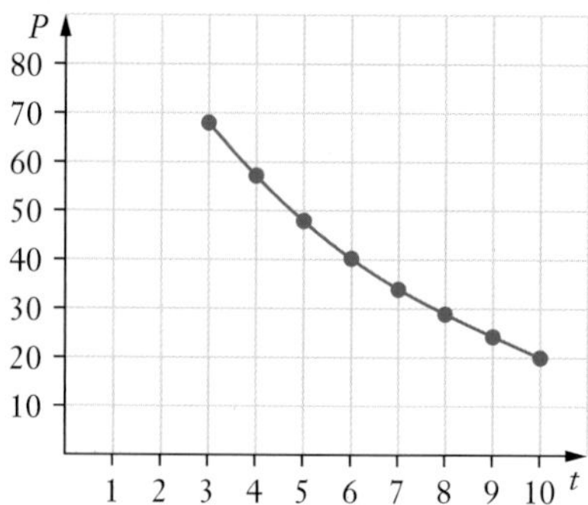

ISBN 9780170390408

8 An environmental group commence a long-term project to reduce the level of pollution in a particular stretch of river.

Starting the campaign in 2004 (when $t = 0$), the pollution level P, in parts per million, is monitored each year and the results are graphed as shown.

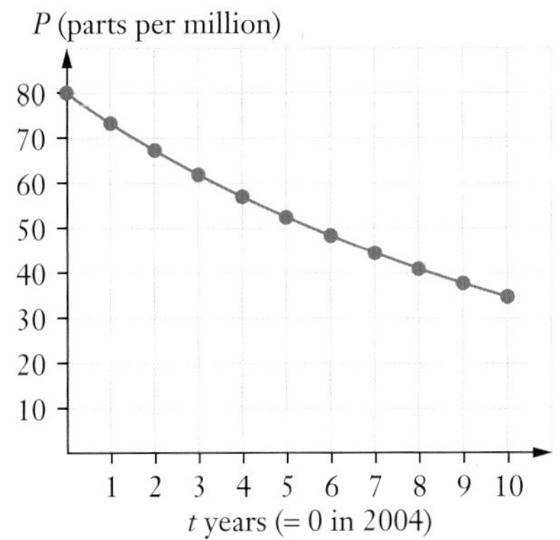

The fall in P is thought to follow an exponential decline according to the rule

$$P = ka^t.$$

a Determine the value of k and of a.

b Use your formula to predict the value of P for 2017.

c The environmental group plan to release a number of fish into the river when they first record a value of P that is less than 20. When is this likely to be?

9 In a particular test area scientists note that when measures are introduced to reduce the population of a particular animal, animal A, classified as a pest, there is a rise in the population of another animal, animal B.

The scientists find that the decrease in the population of animal A and the increase in the population of animal B can both be modelled as exponential growth.

If P_A and P_B are the assessed populations of A and B respectively, then t months after the introduction of control measures the populations are approximately given by:

$$P_A = 10\,000(0.75)^t \qquad \text{and} \qquad P_B = 1000(1.09)^t.$$

Find

a the initial (i.e. $t = 0$) population of A and B in this test area,

b the population of A and B after 3 months of the control program (give answers correct to nearest 50),

c the value of t, correct to one decimal place, when the populations are equal.

10 To control an infestation of a certain flying insect in an area, a number of sterile male insects are released into the area each week.

To monitor the effectiveness of the program, traps are erected at the start of the program and again after 3, 6 and 10 weeks. Each time, the traps are erected at the same time of the day, at the same place and for the same amount of time; the number of these insects caught is then noted. The results are shown in the graph below.

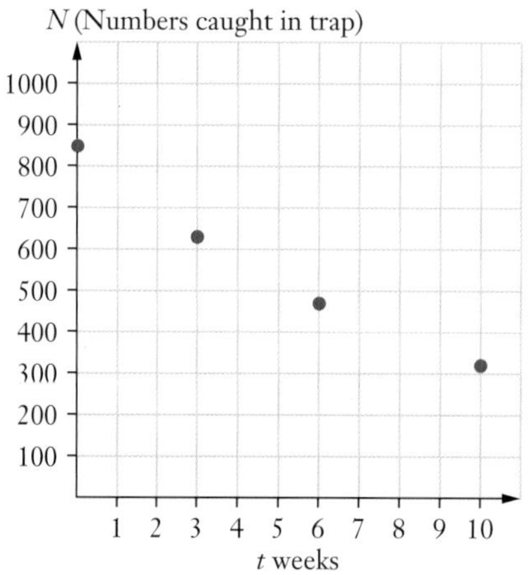

a If the decline in the numbers caught is modelled by an exponential rule of the form $N = ka^t$, determine estimates for the constants k and a.

b The release of the sterile males will cease when the numbers caught in the traps is one quarter of the numbers caught initially (i.e. when $t = 0$).

After how many weeks is this likely to be?

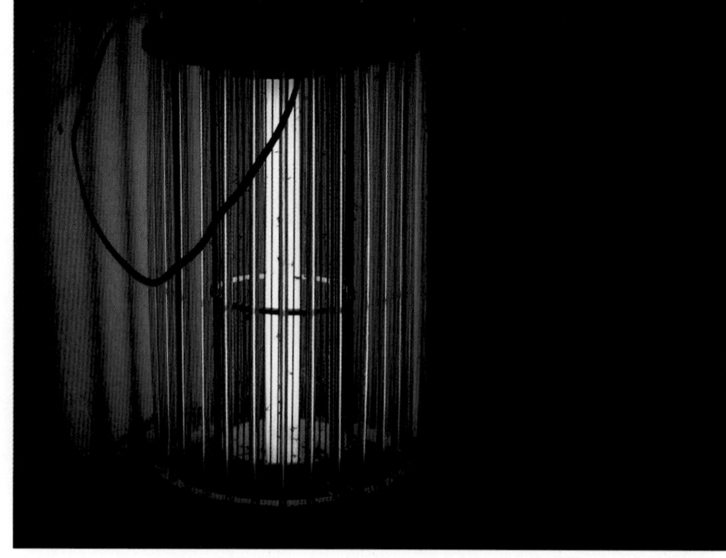

Shutterstock.com/littlekop

ISBN 9780170390408

Miscellaneous exercise two

This miscellaneous exercise may include questions involving the work of this chapter, the work of any previous chapters, and the ideas mentioned in the Preliminary work section at the beginning of the book.

1 For each of the following, state which answer, I, II, III or IV shows the given number written in standard form, or scientific notation, i.e. in the form $A \times 10^n$, where A is a number between 1 and 10 and n is an integer.

		I	II	III	IV
a	36	3.6×10^0	3.6×10^1	3.6×10^2	3.6×10^3
b	0.000 023	3.2×10^{-5}	2.3×10^5	2.3×10^{-4}	2.3×10^{-5}
c	41 000	41×10^3	410×10^2	4.1×10^4	4.1×10^{-4}
d	0.245	2.45×10^{-1}	0.245×10^0	2.45×10^{-2}	2.45×10^2
e	0.003	3×10^{-2}	30×10^{-2}	3×10^{-3}	300
f	912 000	9.12×10^4	9.12×10^7	9.12×10^6	9.12×10^5
g	0.000 002 81	2.81×10^5	2.81×10^{-5}	2.81×10^{-6}	281×10^{-8}
h	14 200 000	1.42×10^7	1.42×10^4	1.42×10^5	1.42×10^6

2 Solve each of the following equations (without the assistance of a calculator).

a $x^2 = 49$ **b** $x^2 = 100$ **c** $x^3 = 1000$

d $2^x = 4$ **e** $3^x = 81$ **f** $5^x + 11 = 12$

g $6^x + 9 = 225$ **h** $4^x = \dfrac{1}{4}$ **i** $4^x = \dfrac{1}{16}$

j $4^x = \dfrac{1}{64}$ **k** $2^x = 0.5$ **l** $2^x = 0.25$

m $2^x = 0.125$ **n** $16x^4 = 400x^2$ **o** $8^{2x+1} = 4^{1-x}$

p $\sqrt{50}x - \sqrt{18}x = \sqrt{2}$ **q** $\sqrt{50x} - \sqrt{18x} = \sqrt{2}$ **r** $(x^3 + 5)(x^3 - 5) = 704$

3 Round each of the following to the number of significant figures stated.

a 12 405 correct to two significant figures.

b 12 607 405 correct to four significant figures.

c 0.000 256 correct to two significant figures.

d 5.63 correct to one significant figure.

e 12 626.8 correct to four significant figures.

4 a What will be the equation of the graph obtained by translating the graph of the function $y = 2^x$ three units to the left? Write your answer both as $y = 2^{f(x)}$ and as $y = k \times 2^x$.

b What will be the equation of the graph obtained by translating the graph of the function $y = 3^x$ two units down?

ISBN 9780170390408

5 The graph below left shows four points which obey the rule $y = 5^x$.

Joining these points with a smooth curve, as shown below right, allows values for other powers of x to be suggested, e.g. $5^{2.8} \approx 91$.

Use the graph to suggest a value for $5^{1.6}$, $5^{2.4}$ and $5^{2.5}$ and then check your answers using a calculator.

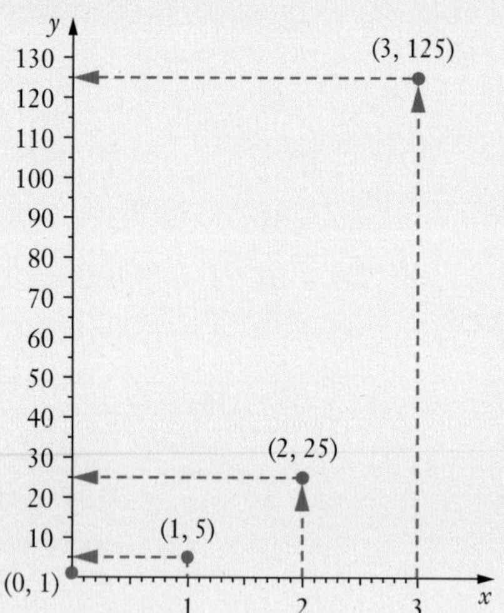

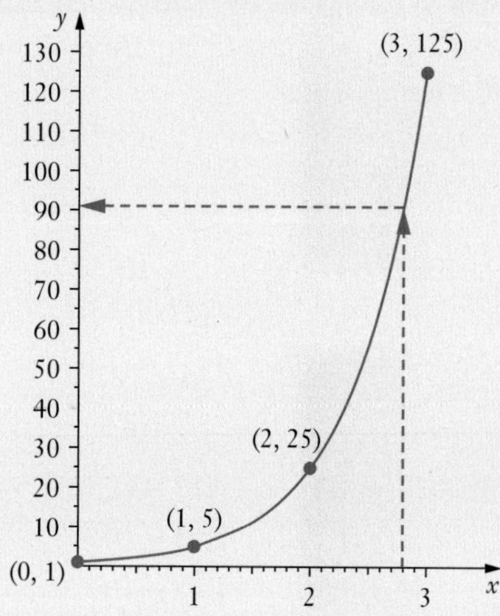

6 Without the assistance of a calculator, solve $(25 \times 5^x - 1)(5^x - 1) = 0$.

7 Without the assistance of a calculator, solve $(2^x)^2 - 5(2^x) + 4 = 0$.

8 A forensic scientist is called to the scene of a murder. Upon arrival at 10 a.m., the scientist notes the temperature of the body as being 23.9°C. This is 18.9°C above the temperature of the surrounding air (which is 5°C). The scientist monitors this 'body temperature above surrounding temperature of 5°C' at half hour intervals. The data collected is shown in the graph shown.

a With T and t as defined in the graph and assuming the relationship between T and t is of the form $T = ka^t$, find the values of the constants k and a.

b If normal body temperature is 37°C, i.e. 32°C above the 5°C temperature of the surroundings, estimate the time of death.

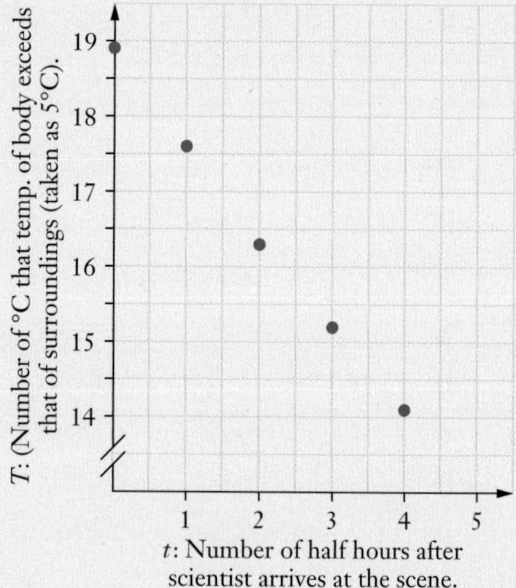

ISBN 9780170390408

Sequences

- Sequences
- Arithmetic sequences
- Geometric sequences
- Jumping to later terms of arithmetic and geometric sequences
- Growth and decay – again!
- Miscellaneous exercise three

Situation One

Rabbits

The Fibonacci sequence 1, 1, 2, 3, 5, 8, 13, … is named after the Italian mathematician Leonardo Fibonacci (1170 – 1240). The sequence occurs in many number patterns associated with nature. One example involving rabbits is shown here. Commencing with one pair of rabbits, we assume that these rabbits will be adults after 1 month and will produce a pair of baby rabbits each month after that.

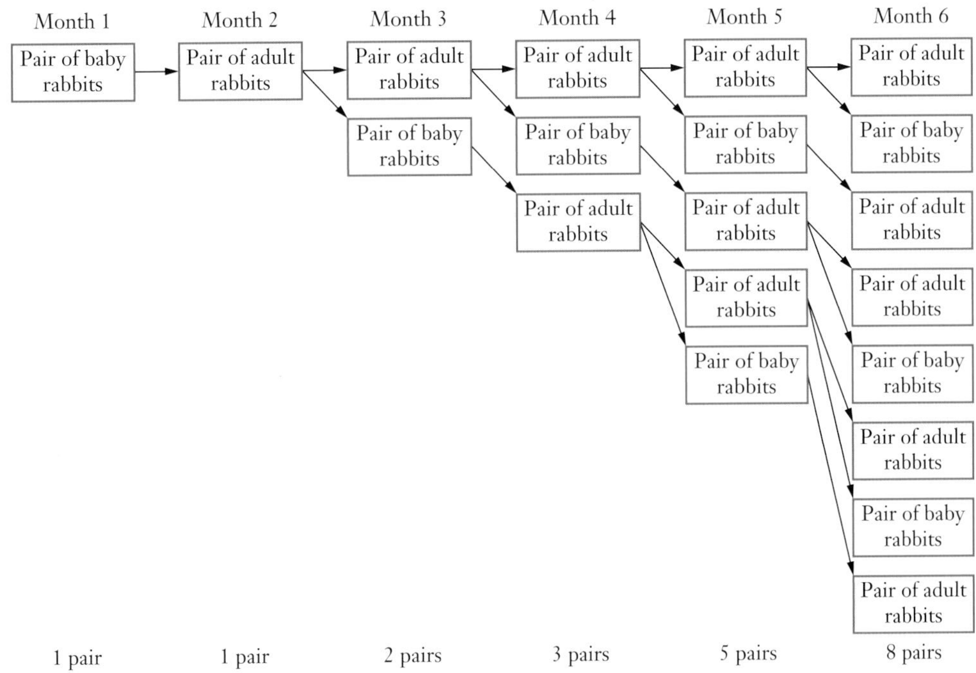

Notice that in the number sequence 1, 1, 2, 3, 5, 8, 13, …
each number, after the first two ones, is the sum of the previous two numbers.

$1 + 1 = 2$

$1 + 2 = 3$

$2 + 3 = 5$

$3 + 5 = 8$

$5 + 8 = 13$

Shutterstock.com/americ 181

a Write the next five numbers in this sequence.

b Use the internet to investigate real world applications of the Fibonacci sequence of numbers.

Situation Two

Hexagonal patchworks

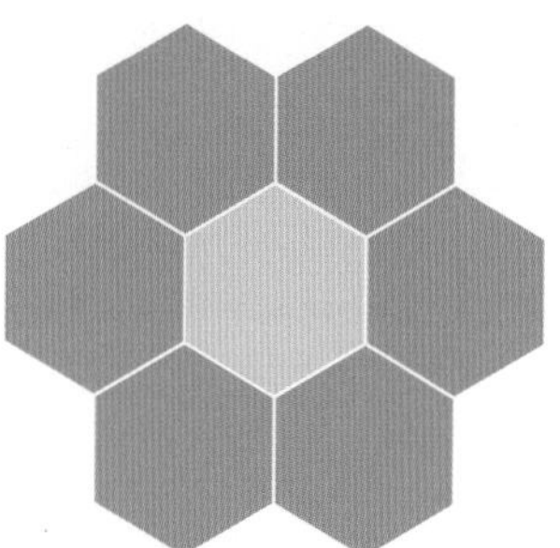

Patchwork one shown on the right has a central hexagon and a ring of hexagons around it.

Placing a second ring of hexagons around Patchwork one gives *Patchwork two* shown below left and a further ring of hexagons around Patchwork two gives *Patchwork three* shown below right.

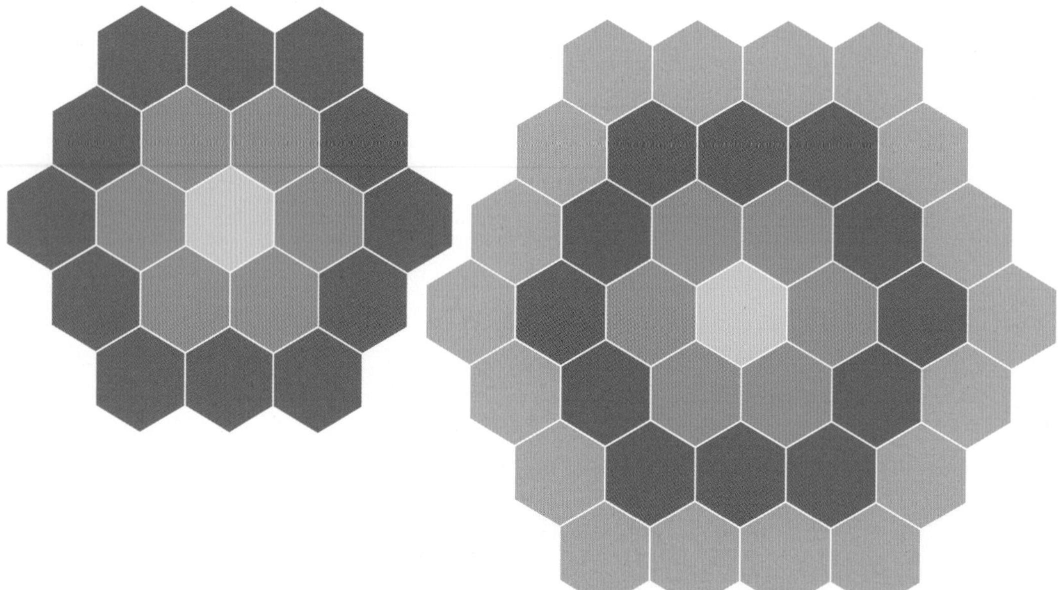

For the three patchworks shown above, consider things like:

- the number of hexagons in the outer ring of each patchwork
- the number of hexagonal pieces in each patchwork
- the number of edges that form the border of each patchwork
- other aspects you may think of

and use your answers to try to predict what numbers the next few patchworks in the continuing pattern would have for these things.

Shutterstock.com/Maria Dryfhout

ISBN 9780170390408

Sequences

WS Classifying sequences

Consider the following table of values for the linear function $y = 3x + 2$.

x	2	3	4	5	6	7	8
y	8	11	14	17	20	23	26

All pairs (x, y) in the table fit the rule $y = 3x + 2$

For example, for (2, 8) $8 = 3(2) + 2$
for (3, 11) $11 = 3(3) + 2$
for (4, 14) $14 = 3(4) + 2$ etc.

However let us now just consider the **sequence** of y-values, i.e.

8 11 14 17 20 23 26

Note

- A **sequence** is a set of items belonging in a certain order, according to some rule. Knowing sufficient items, and the rule that is involved, the next item can be determined.
- In this course we will concentrate on *number* sequences.
- We refer to each number in the sequence as a **term** of the sequence. Thus, in the above sequence the first term is 8, the second term is 11, the third term is 14, the fourth term is 17 and so on.
- Writing T_1 for the first term, T_2 for the second term and so on, we have:

  $T_1 = 8$ $T_2 = 11$ $T_3 = 14$ $T_4 = 17$ $T_5 = 20$ $T_6 = 23$ $T_7 = 26$
- The situations at the beginning of this chapter involved sequences of numbers, for example the Fibonacci sequence:

  1 1 2 3 5 8 13 ...
- Whilst we usually use $T_1, T_2, T_3, T_4, T_5, \ldots$ (or perhaps $t_1, t_2, t_3, t_4, t_5, \ldots$) for the terms of a sequence, other letters may be used at times. For example, for the terms of the Fibonacci sequence, we might use $F_1, F_2, F_3, F_4, F_5, \ldots$ Some calculators may use U or a, or some other letter.

EXAMPLE 1

For the sequence 3, 10, 31, 94, 283, 850, 2551, ... determine

a T_3 **b** T_5 **c** $T_3 + T_5$ **d** $3T_2$ **e** $2T_3$ **f** T_8

Solution

a T_3 is the 3rd term in the sequence. Thus $T_3 = 31$.

b T_5 is the 5th term in the sequence. Thus $T_5 = 283$.

c $T_3 + T_5 = 31 + 283$
$= 314$

d $3T_2 = 3(10)$
$= 30$

e $2T_3 = 2(31)$
$= 62$

f Noticing that each term is obtained by multiplying the previous term by 3 and then adding 1, it follows that $T_8 = 3(T_7) + 1$
$= 3(2551) + 1$
$= 7654$.

ISBN 9780170390408

Exercise 3A

For the sequence 10, 14, 18, 22, 26, 30, 34, 38, … determine

1 T_3 **2** T_5 **3** $T_3 + T_5$ **4** T_8

5 $3T_2$ **6** $2T_3$ **7** $3(T_1 + T_2)$ **8** $3T_1 + T_2$

9 T_9 **10** T_{10} **11** $(T_3)^2$ **12** $(T_2)^3$

For the sequence 5, 8, 11, 14, 17, 20, 23, … determine

13 T_2 **14** T_6 **15** $T_2 + T_6$ **16** T_8

17 T_9 **18** $T_3 + 2T_1$ **19** $T_1 + 2T_3$ **20** $(T_3 - T_2)^2$

For the sequence 2, 6, 18, 54, 162, 486, 1458, … determine

21 T_5 **22** $3T_2$ **23** $T_1 + T_2 + T_3$ **24** T_8

Using C_n for the nth term of the cubic numbers 1, 8, 27, 64, 125, … determine

25 C_3 **26** C_6 **27** C_7 **28** $C_6 - C_5$

The Lucas sequence follows the same rule as the Fibonacci sequence, i.e. each term after the first two is the sum of the previous two terms. Using L_n for the nth term of the Lucas sequence and given that $L_1 = 1$ and $L_2 = 3$, determine

29 L_3 **30** L_4 **31** $(L_4)^2$ **32** $2L_8$

Arithmetic sequences

Arithmetic progressions

WS

Arithmetic sequences

x	1	2	3	4	5	6
y	1	3	5	7	9	11

Notice that in the table of values on the right, as the x-values increase by 1 the y-values increase by 2. As we would expect from this constant first difference pattern of 2, graphing gives points that lie in a **straight line** and the gradient of the line is 2.

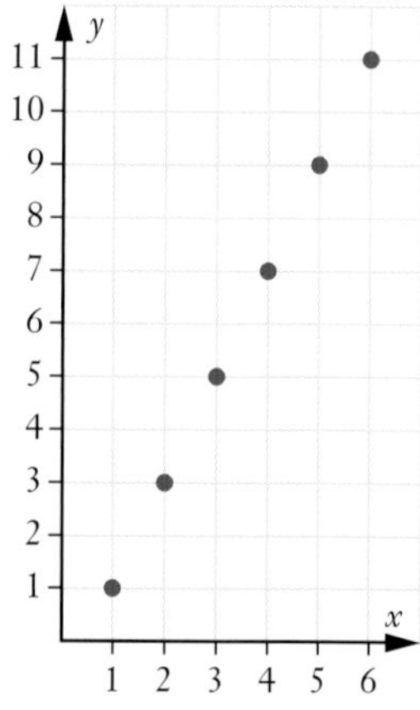

Sequences of numbers in which each term is obtained from the previous term by the addition of some constant number are said to be

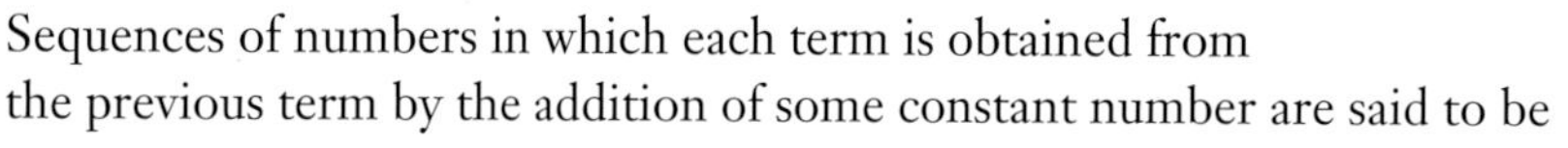

Arithmetic Sequences,

Arithmetic Progressions

or simply *APs.*

For example 1, 2, 3, 4, 5, 6, …

In this case we say that **the first term** is 1 and **the common difference** is 1.

Similarly for the AP: 1, 3, 5, 7, 9, 11, …

we say that the first term is 1 and the common difference is 2.

For 7, 11, 15, 19, 23, 27, …

the first term is 7 and the common difference is 4.

For 67, 62, 57, 52, 47, 42, …

the first term is 67 and the common difference is –5.

ISBN 9780170390408

Thus all arithmetic sequences are of the form:

$$a, \quad a+d, \quad a+2d, \quad a+3d, \quad a+4d, \quad a+5d, \quad a+6d, \quad \ldots$$

In this general form we have a first term of 'a' and common difference 'd'.

Consider again the arithmetic sequence:

$$1, \quad 3, \quad 5, \quad 7, \quad 9, \quad 11, \quad \ldots$$

Each term is obtained by adding 2 to the previous term. Thus, if T_n is the nth term, it follows that the next term, T_{n+1}, will be $T_n + 2$.

Hence, for this sequence $\quad T_{n+1} = T_n + 2 \quad$ (or $\quad t_{n+1} = t_n + 2$).

For the general arithmetic sequence at the top of the page

$$T_{n+1} = T_n + d \quad \text{(or} \quad t_{n+1} = t_n + d\text{)}.$$

This rule tells us how the terms of the sequence *recur*. It is the **recursive rule** or **recursive equation** for the sequence. Given this rule and one term, usually the first, all other terms of the sequence can be determined.

EXAMPLE 2

A sequence is such that $T_{n+1} = T_n + 5$ and the first term, T_1, is 7. Find the first four terms of the sequence.

Solution

The recursive definition informs us that each term is the previous term add 5.

Hence, if $T_1 = 7$,

it follows that $\quad T_2 = 7 + 5 = 12 \qquad T_3 = 12 + 5 = 17 \qquad T_4 = 17 + 5 = 22$

The first four terms of the sequence are 7, 12, 17, 22.

EXAMPLE 3

For each of the following sequences, state whether the sequence is an AP or not and, for those that are, state the first term, the common difference and a recursive formula.

a 13, 18, 23, 28, 33, 38, ...,

b 3, 6, 12, 24, 48, 96, ...,

c 90, 79, 68, 57, 46, 35, ...,

Solution

a Each term is 5 more than the previous term. Thus the sequence *is* an arithmetic progression.

First term = 13

Common difference = 5

Recursive formula: $T_{n+1} = T_n + 5$

b The terms do not have a common difference. Thus the sequence is *not* an arithmetic progression.

c Each term is 11 less than the previous term. Thus the sequence is an arithmetic progression.

First term = 90

Common difference = −11

Recursive formula: $T_{n+1} = T_n - 11$

Note: The recursive rule from part **a** in the previous example, i.e.

$$T_{n+1} = T_n + 5$$

could equally well be written in the form

$$T_n = T_{n-1} + 5.$$

Both expressions tell us the same thing, i.e. that each term of the sequence is obtained by adding 5 to the previous term.

EXAMPLE 4

A sequence is defined by $T_n = 3T_{n-1} - 2$ with $T_1 = 5$. Determine the first five terms of this sequence and hence determine whether the sequence is an arithmetic sequence or not.

Solution

The formula $T_n = 3T_{n-1} - 2$ tells us that each term is obtained by multiplying the previous term by 3 and then subtracting 2.

Thus, if $T_1 = 5$, it follows that

$$T_2 = 3(5) - 2 \quad = 13,$$
$$T_3 = 3(13) - 2 \quad = 37,$$
$$T_4 = 3(37) - 2 \quad = 109,$$
$$T_5 = 3(109) - 2 \quad = 325.$$

The first five terms are 5, 13, 37, 109, 325.

These terms do *not* have a common difference.
The sequence is *not* an arithmetic sequence.

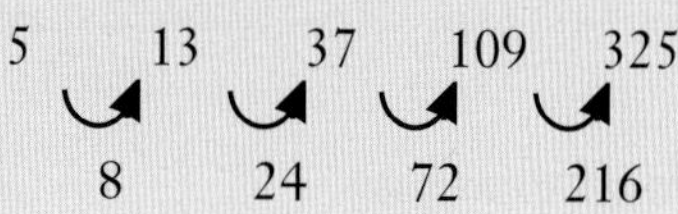

Alternatively, for the previous example, a calculator could be used to display the terms of the sequence, once an appropriate recursive rule and first term have been entered.

☑ $a_{n+1} = 3 \cdot a_n - 2$
$a_1 = 5$
☐ $b_{n+1} = \square$
$b_1 = 0$
☐ $c_{n+1} = \square$
$c_1 = 0$

n	a_n
1	5
2	13
3	37
4	109
5	325

ISBN 9780170390408

Notice that the display on the right is similar to that shown on the previous page but now the progressive sums (also called partial sums) of the terms are also displayed.

$T_1 = 5$

$T_1 + T_2 = 5 + 13 = 18$

$T_1 + T_2 + T_3 = 5 + 13 + 37 = 55$

$T_1 + T_2 + T_3 + T_4 = 5 + 13 + 37 + 109 = 164$

$T_1 + T_2 + T_3 + T_4 + T_5 = 5 + 13 + 37 + 109 + 325 = 489$

☑	$a_{n+1} = 3 \cdot a_n - 2$
	$a_1 = 5$
☐	$b_{n+1} = \square$
	$b_1 = 0$
☐	$c_{n+1} = \square$
	$c_1 = 0$

n	an	Σa_n
1	5	5
2	13	18
3	37	55
4	109	164
5	325	489

Calculators differ in the way they accept and display such information. If you wish to use a calculator in this way, make sure that **you** can use **your** calculator to input recursive formulae and to display the terms of a sequence.

Spreadsheets are another way of displaying the terms of a recursively defined sequence as shown below for the sequence with recursive definition

$$T_{n+1} = 2T_n + 1 \text{ with } T_1 = 3$$

	A	B	C	D
1	3	3		
2	7	10		
3	15	25		
4	31	56		
5	63	119		
6	127	246		
7	255	501		
8	511	1012		
9	1023	2035		
10	2047	4082		

= 2*A3 + 1

= Sum (A$1:A6)

What does the inclusion of the $ symbol do?

Note that this sequence is **not** an arithmetic sequence – the entries in column A do not display a common difference pattern.

Once again, the progressive sums (or partial sums) can easily be shown, as in column B.

Create the spreadsheet yourself and use the 'fill down' ability when creating it.

ISBN 9780170390408

Geometric progressions 1

Geometric progressions 2

Geometric sequences

Geometric sequences

x	1	2	3	4	5	6
y	3	6	12	24	48	96

Notice that in the table of values on the right, as the x-values increase by 1 the y-values multiply by 2. As we would expect from this constant ratio of successive y-values, graphing gives the characteristic shape of an **exponential** function.

Sequences that progress by each term being obtained by multiplying the previous term by a constant number are said to be

Geometric sequences,

Geometric progressions,

or simply *GPs.*

For example, 5, 15, 45, 135, 405, 1215, …

In this case we say that **the first term** is 5 and **the common ratio** is 3.

Similarly, for the GP: 0.5, 1, 2, 4, 8, 16, …

we say that the first term is 0.5 and the common ratio is 2.

For 1000, 100, 10, 1, 0.1, 0.01, …

the first term is 1000 and the common ratio is 0.1.

For 64, 96, 144, 216, 324, 486, …

the first term is 64 and the common ratio is 1.5.

Thus all GPs are of the form:

$$a, \quad ar, \quad ar^2, \quad ar^3, \quad ar^4, \quad ar^5, \quad ar^6, \quad \ldots$$

In this general form we have a first term of 'a' and common ratio 'r'.

Using recursive notation, we have $T_{n+1} = r \times T_n$ with $T_1 = a$, (or $t_{n+1} = rt_n$, $t_1 = a$).

EXAMPLE 5

For each of the following sequences state whether the sequence is a geometric sequence or not and, for those that are, state the first term, the common ratio and a recursive formula.

a 3, 6, 12, 24, 48, 96, …,

b 128, 96, 72, 54, 40.5, 30.375, …,

c 4, 9, 14, 19, 24, 29, …,

Solution

a $\frac{6}{3} = 2, \quad \frac{12}{6} = 2, \quad \frac{24}{12} = 2, \quad \frac{48}{24} = 2, \quad \frac{96}{48} = 2.$

Each term is the previous term multiplied by 2.

Thus the sequence *is* a geometric sequence.

First term = 3

Common ratio = 2

Recursive formula: $T_{n+1} = 2T_n$

(or $T_n = 2T_{n-1}$)

ISBN 9780170390408

b $\frac{96}{128} = 0.75$, $\frac{72}{96} = 0.75$, $\frac{54}{72} = 0.75$, $\frac{40.5}{54} = 0.75$, $\frac{30.375}{40.5} = 0.75$.

Each term is the previous term multiplied by 0.75.
Thus the sequence *is* a geometric sequence.

First term = 128

Common ratio = 0.75

Recursive formula: $T_n = 0.75T_{n-1}$

c $\frac{9}{4} = 2.25$, $\frac{14}{9} = 1.\bar{5}$.

The sequence does not have a common ratio.
Thus the sequence is *not* a geometric sequence.

EXAMPLE 6

$400 is invested in an account and earns $20 interest each year.

a How much is the account worth after 1 year, 2 years, 3 years and 4 years?

b Do the amounts the account is worth at the end of each year form an arithmetic sequence, a geometric sequence or neither of these?

Solution

a

Initial value	Value after 1 year	Value after 2 years	Value after 3 years	Value after 4 years	...
$400	$400 + 1($20)	$400 + 2($20)	$400 + 3($20)	$400 + 4($20)	...
$400	$420	$440	$460	$480	...

After 1, 2, 3 and 4 years the account is worth $420, $440, $460 and $480 respectively.

b The situation gives rise to amounts with a common difference of $20.
The amounts the account is worth at the end of each year form an arithmetic sequence.

As you are probably aware, the situation described in the previous example is not the way that an investment usually earns interest. Once the $20 interest has been added at the end of year 1, the account has $420 in it and it is this $420 that attracts interest in year 2, not just the initial $400. In this way, the interest earned in one year itself attracts interest in subsequent years, i.e. compound interest is involved, rather than the simple interest situation described in the previous example.

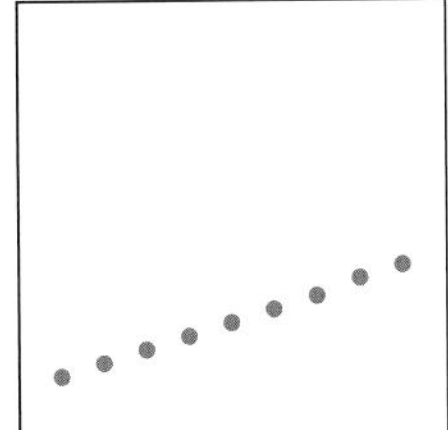

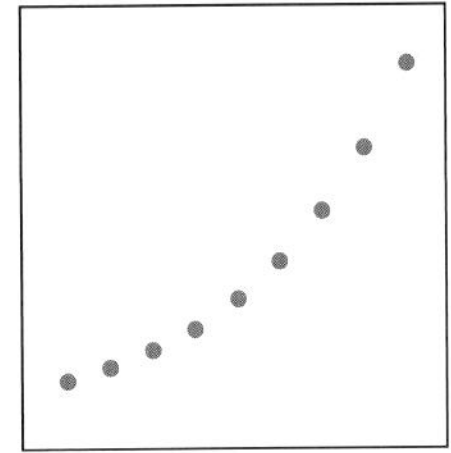

ISBN 9780170390408

EXAMPLE 7

\$2000 is invested and accrues interest at a rate of 10% per annum, compounded annually.

a If no further deposits are made, how much will be in the account after 1 year, 2 years, 3 years and 4 years?

b Do the amounts the account is worth at the end of each year form an arithmetic sequence, a geometric sequence or neither of these?

Solution

a

Initial value	Value after 1 year	Value after 2 years	Value after 3 years	Value after 4 years	...
\$2000	$\$2000 \times 1.1$	$\$2000 \times 1.1^2$	$\$2000 \times 1.1^3$	$\$2000 \times 1.1^4$	...
\$2000	\$2200	\$2420	\$2662	\$2928.20	...

After 1, 2, 3 and 4 years the account is worth \$2200, \$2420, \$2662 and \$2928.20 respectively.

b The situation gives rise to amounts with a common ratio of 1.1.
The amounts the account is worth at the end of each year form a geometric sequence.

Exercise 3B

For each of the following arithmetic sequences state

- the first term, T_1

and

- the $(n + 1)$th term, T_{n+1}, in terms of the nth term, T_n.

1 6, 10, 14, 18, 22, 26, ...

2 28, 26, 24, 22, 20, 18, ...

3 5, 15, 25, 35, 45, 55, ...

4 7.5, 10, 12.5, 15, 17.5, 20, ...

5 100, 89, 78, 67, 56, 45, ...

For each of the following geometric sequences state

- the first term, T_1,

and

- the nth term, T_n, in terms of the $(n - 1)$th term, T_{n-1}.

6 6, 12, 24, 48, 96, 192, ...

7 0.375, 1.5, 6, 24, 96, 384, ...

8 384, 96, 24, 6, 1.5, 0.375, ...

9 50, 150, 450, 1350, 4050, 12 150, ...

10 1000, 1100, 1210, 1331, 1464.1, 1610.51, ...

ISBN 9780170390408

For each of the following sequences, state whether the sequence is arithmetic, geometric or neither of these two types.

11	2,	5,	11,	23,	47,	95,	…
12	1,	5,	25,	125,	625,	3125,	…
13	13,	14.5,	16,	17.5,	19,	20.5,	…
14	50,	39,	28,	17,	6,	–5,	…
15	1,	1,	2,	3,	5,	8,	…
16	128,	160,	200,	250,	312.5,	390.625,	…

17 $T_{n+1} = 3T_n, T_1 = 3.$

18 $T_{n+1} = T_n + 6, T_1 = 2.$

19 $T_{n+1} = 3T_n + 5, T_1 = 1.$

20 $T_n = (T_{n-1})^2, T_1 = 7.$

21 $T_n = T_{n-1} - 8, T_1 = 2000.$

22 $T_n = (0.5)T_{n-1}, T_1 = 8.$

23 An AP has a first term of 8 and a common difference of 3. Determine the first four terms of the sequence and the recursive rule for T_{n+1} in terms of T_n.

24 An AP has a first term of 100 and a common difference of –3. Determine the first four terms of the sequence and the recursive rule for T_{n+1} in terms of T_n.

25 A GP has a first term of 11 and a common ratio of 2. Determine the first four terms of the sequence and the recursive rule for T_{n+1} in terms of T_n.

26 A GP has a first term of 2048 and a common ratio of 0.5. Determine the first four terms of the sequence and the recursive rule for T_{n+1} in terms of T_n.

27 The graph on the right shows the number of vehicles a company sold in a particular country each year from 2011 to 2014.

a Verify that the figures for these years are in arithmetic progression.

b With $N_{2011} = 6000$, write a recursive rule for N_{n+1} in terms of N_n.

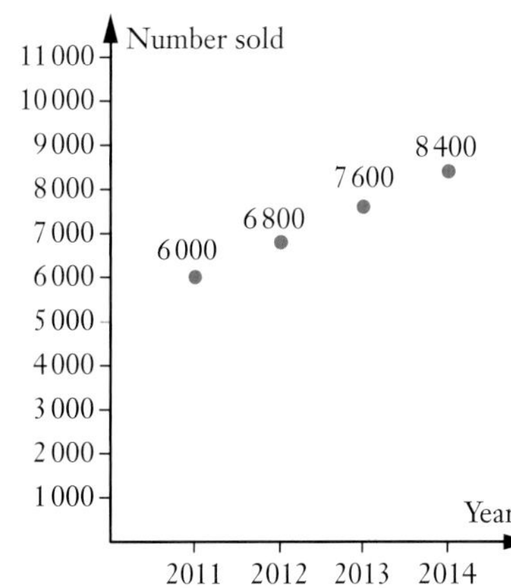

ISBN 9780170390408

28 Each term of a sequence is obtained using the recursive rule

$$T_n = T_{n-1} + 10\% \text{ of } T_{n-1}$$

a Is the sequence an arithmetic progression, a geometric progression or neither of these?

b If the first term of the sequence is 500, find the next three terms.

29 Each term of a sequence is obtained using the recursive rule

$$T_n = T_{n-1} + 25\% \text{ of } T_{n-1}$$

a Is the sequence an arithmetic progression, a geometric progression or neither of these?

b If the first term of the sequence is 1000, find the next three terms.

30 Each term of a sequence is obtained using the recursive rule

$$T_n = T_{n-1} - 10\% \text{ of } T_{n-1}$$

a Is the sequence an arithmetic progression, a geometric progression or neither of these?

b If the first term of the sequence is 24 000, find the next three terms.

31 The graph on the right indicates the first five terms, T_1 to T_5, of a sequence, all of which are whole numbers.

State **a** the first term and a recursive rule for the sequence,

b whether the sequence is arithmetic, geometric or neither of these types.

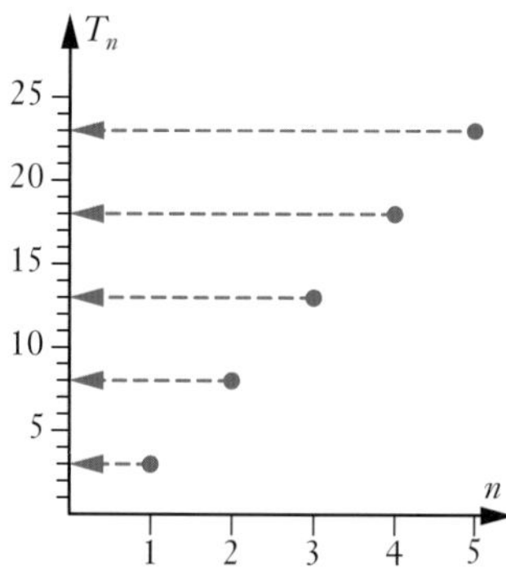

32 The graph on the right indicates the first five terms, T_1 to T_5, of a sequence. The last four of these terms are whole numbers.

State **a** the first term and a recursive rule for the sequence,

b whether the sequence is arithmetic, geometric or neither of these types.

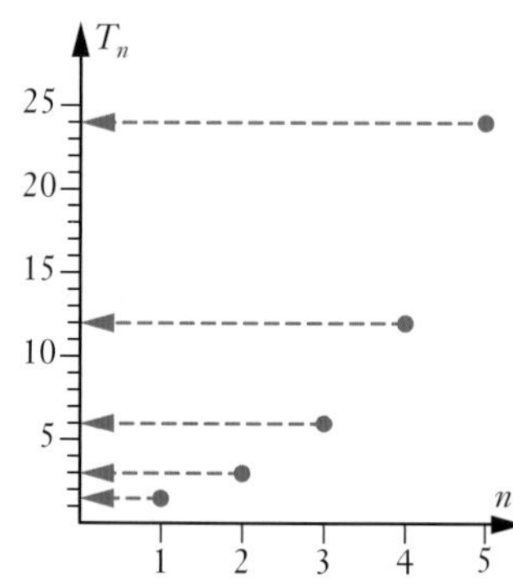

33 The graph on the right indicates the first five terms, T_1 to T_5, of a sequence. All of these terms are whole numbers.

Determine **a** the first five terms,

b whether the sequence is arithmetic, geometric or neither of these types.

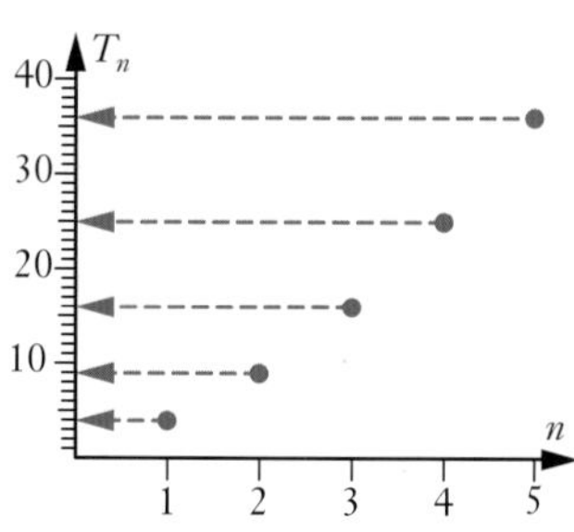

ISBN 9780170390408

34 \$1200 is invested in an account and earns \$96 interest each year.

- **a** How much is the account worth after 1 year, 2 years, 3 years and 4 years?
- **b** Are these amounts an arithmetic progression, geometric progression or neither of these?
- **c** Express the sequence of values:

  Initial value, value after 1 year, value after 2 years, ...

  using recursive notation.

35 The number of stitches in each row of a particular crochet pattern are as shown on the right.

1st row $\oplus \; \oplus \; \oplus \; \oplus$
2nd row $\oplus \; \oplus \; \oplus \; \oplus \; \oplus$
3rd row $\oplus \; \oplus \; \oplus \; \oplus \; \oplus \; \oplus$
4th row $\oplus \; \oplus \; \oplus \; \oplus \; \oplus \; \oplus \; \oplus$

With T_1 the number of stitches in the first row, T_2 the number in the second row, etc., express the sequence

$$T_1, \quad T_2, \quad T_3, \quad T_4, \quad T_5, \quad \ldots$$

using recursive notation (i.e., state T_1 and the recursive rule).

36 Won Yim starts working for a particular company on the 1st January one year and is paid an initial annual salary of \$45 000 with a guaranteed \$1500 rise each year for the next 7 years.

Express the sequence of annual salaries over this time as a sequence using recursive notation and state whether the terms of the sequence progress arithmetically, geometrically or neither of these.

37 Joe started a new job on 1 January 2014 and, during 2014, he received a salary of \$68 000. His contract guarantees a salary increase of 5% of the salary of the previous year on each subsequent 1 January, until and including 1 January 2017. Calculate Joe's salary for each year from 2014 to 2017.

Express the sequence of salaries from 2014 (term one) to 2017 (term four) using recursive notation.

38 \$1500 is invested and accrues interest at a rate of 8% per annum, compounded annually. With this \$1500 as the first term in the sequence, express the value of the account on this and each subsequent year as a sequence defined recursively.

39 Each year the value of a car depreciates by 15% of its value at the beginning of that year. The car is initially worth \$36 000. With this \$36 000 as the first term, express the value of the car on this and each subsequent year as a sequence defined recursively.

Shutterstock.com/G-Valeriy

Jumping to later terms of arithmetic and geometric sequences

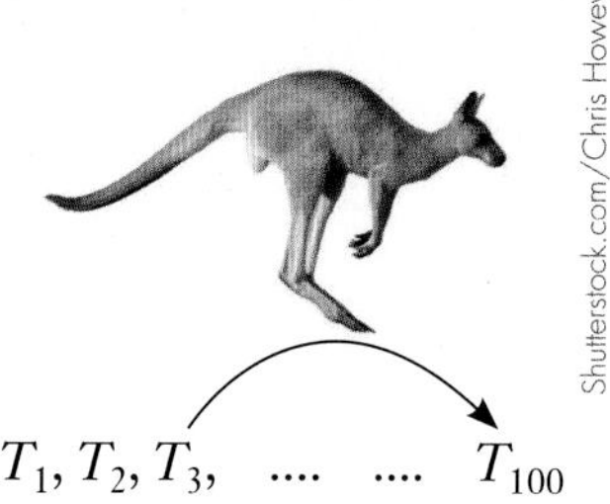

Shutterstock.com/Chris Howey

Consider the arithmetic sequence defined by

$$T_{n+1} = T_n + 2 \text{ and } T_1 = 3$$

The rule allows us to obtain the terms of the sequence:

$$T_1 = 3$$
$$T_2 = T_1 + 2 = 3 + 2 = 5$$
$$T_3 = T_2 + 2 = 5 + 2 = 7$$
$$T_4 = T_3 + 2 = 7 + 2 = 9 \qquad \text{etc.}$$

However, if we wanted to know the value of a term much later in the sequence, say T_{100}, it would be a tedious process to have to calculate all of the terms up to T_{100}. It would be more useful if we could jump to the desired term without having to determine all of the preceding ones.

EXAMPLE 8

For the sequence defined recursively as $T_{n+1} = T_n + 7$ with $T_1 = 25$, determine the first four terms and the one hundredth term.

Solution

With $T_{n+1} = T_n + 7$, it follows that

$$\begin{aligned} T_2 &= T_1 + 7 \\ &= 32 \end{aligned}$$

$$\begin{aligned} T_3 &= T_2 + 7 \\ &= 39 \end{aligned}$$

$$\begin{aligned} T_4 &= T_3 + 7 \\ &= 46 \end{aligned}$$

Notice that by the second term we have added 7 *once*, by the third term we have added 7 *twice*, by the fourth term we have added 7 *three* times. It follows that for the one hundredth term we need to have added 7 ninety-nine times.

Hence
$$\begin{aligned} T_{100} &= T_1 + 99(7) \\ &= 25 + 693 \\ &= 718. \end{aligned}$$

The first four terms are 25, 32, 39 and 46 and the one hundredth term is 718.

ISBN 9780170390408

If we apply the thinking used in the previous example to the general arithmetic sequence

$$a, \quad a+d, \quad a+2d, \quad a+3d, \quad a+4d, \quad a+5d, \quad \ldots$$

we note that for T_2 the common difference, d, has been added once, for T_3 it has been added twice, for T_4 it has been added three times, and so on. Thus, for T_n we need to add the common difference $(n-1)$ times.

Thus, the arithmetic progression

$$a, \quad a+d, \quad a+2d, \quad a+3d, \quad a+4d, \quad a+5d, \quad \ldots$$

has an nth term given by:

$$T_n = a + (n-1)d \qquad \text{or} \qquad t_n = t_1 + (n-1)d$$

Just pause for a moment and check that you understand the difference between a recursive formula, which tells you how each term is obtained from the previous term, and the formula for the nth term, which allows you to determine any term.

Note: Thinking of T_n as y, and n as x, the reader should see similarities between

$$T_n = a + (n-1)d \qquad \text{and} \qquad y = mx + c.$$

This is no surprise given the linear nature of arithmetic sequences.

EXAMPLE 9

For the AP: 11, 14, 17, 20, ...

Determine **a** T_{123}

b T_{500}

c which term of the sequence is the first to exceed 1 000 000.

Solution

a $T_{123} = 11 + 122(3)$
$= 377$

b $T_{500} = 11 + 499(3)$
$= 1508$

c Suppose that T_n is the first term to exceed 1 000 000.

Now $T_n = 11 + (n-1)3$

$\therefore \quad 11 + (n-1)3 > 1\,000\,000$

i.e., $n > 333\,330.\overline{6}$

Thus, the first term to exceed 1 000 000 is $T_{333\,331}$.

ISBN 9780170390408

EXAMPLE 10

An AP has a 50th term of 209 and a 61st term of 253. Find

a the 62nd term **b** the 1st term.

Solution

a To go from T_{50} to T_{61} we must add the common difference 11 times.

Thus, if d is the common difference, then $253 - 209 = 11d$

$\therefore$ $44 = 11d$

giving $d = 4$

Hence $T_{62} = T_{61} + 4$

$= 257$

The 62nd term is 257.

b From our understanding of APs, it follows that $T_{50} = T_1 + 49d$

$\therefore$ $209 = T_1 + 49(4)$

$209 - 196 = T_1$

$T_1 = 13$

The 1st term is 13.

Note

Alternatively, we could use the given information to write

$$a + 49d = 209$$

and $$a + 60d = 253$$

and then solve these equations simultaneously.

EXAMPLE 11

For the sequence defined recursively as $T_{n+1} = 1.5T_n$ with $T_1 = 8192$, determine the first four terms and the fifteenth term.

Solution

With $T_{n+1} = 1.5T_n$, it follows that

$$T_2 = 1.5T_1 = 8192 \times 1.5 = 12\,288$$

$$T_3 = 1.5T_2 = 12\,288 \times 1.5 = 18\,432$$

$$T_4 = 1.5T_3 = 18\,432 \times 1.5 = 27\,648$$

Note that for the second term we multiply by 1.5 *once*, for the third term we multiply by 1.5 *twice*, for the fourth term we multiply by 1.5 *three* times. It follows that for the fifteenth term we need to multiply by 1.5 fourteen times.

Hence $T_{15} = T_1 \times 1.5^{14}$

$= 8192 \times 1.5^{14}$

$= 2\,391\,484.5$

The first four terms are 8192, 12 288, 18 432 and 27 648. The fifteenth term is 2 391 484.5.

ISBN 9780170390408

Make sure you can obtain this same value for the fifteenth term of the sequence of the previous example by using a calculator to display terms of the sequence.

Are the larger numbers displayed as shown on the right or does you calculator use scientific notation to display these numbers?

☑ $a_{n+1} = 1.5a_n$
$a_1 = 8192$
☐ $b_{n+1} = \square$
$b_1 = 0$
☐ $c_{n+1} = \square$
$c_1 = 0$

n	a_n
11	472392
12	708588
13	1062882
14	1594323
15	2391484.5

If we apply the thinking of the previous example to the general geometric sequence:

$$a, \quad ar, \quad ar^2, \quad ar^3, \quad ar^4, \quad ar^5, \quad ar^6, \quad \ldots$$

we note that T_2 is ar^1, T_3 is ar^2, T_4 is ar^3, etc. Thus, $T_n = ar^{n-1}$.

Thus the geometric progression

$$a, \quad ar, \quad ar^2, \quad ar^3, \quad ar^4, \quad ar^5, \quad ar^6, \quad \ldots$$

has an nth term given by:

$$T_n = a \times r^{n-1} \qquad \text{or} \qquad t_n = t_1 r^{n-1}$$

Note: Thinking of T_n as y, and n as x, the reader should see similarities between

$$T_n = a \times r^{n-1} \qquad \text{and} \qquad y = k \times b^x.$$

Again, no surprise given the exponential nature of geometric sequences.

EXAMPLE 12

Determine the 12th term and the 15th term of the geometric sequence:

$$0.0025, \quad 0.01, \quad 0.04, \quad 0.16, \quad \ldots$$

Solution

By inspection, the common ratio is 4.

Hence the 12th term will be 0.0025×4^{11}
$= 10\,485.76$

and the 15th term will be 0.0025×4^{14}
$= 671\,088.64$

Again, make sure that you can obtain these same answers using the ability of some calculators to display the terms of a sequence.

EXAMPLE 13

The 13th term of a GP is 12 288 and the 16th term is 98 304. Find

a the 17th term **b** the 1st term.

Solution

a To go from the 13th term to the 16th term we must multiply by the common ratio 3 times.

If r is the common ratio, then $T_{16} = T_{13} \times r^3$

$\therefore$ $98\,304 = 12\,288 \times r^3$

Giving $r = 2$

Hence $T_{17} = T_{16} \times 2$

$= 196\,608$

The 17th term is 196 608.

b From our understanding of GPs, it follows that $T_{13} = T_1 \times r^{12}$

i.e. $12\,288 = T_1 \times 2^{12}$

Giving $T_1 = 3$

The 1st term is 3.

(Alternatively, we could write $ar^{12} = 12\,288$ and $ar^{15} = 98\,304$ and solve simultaneously.)

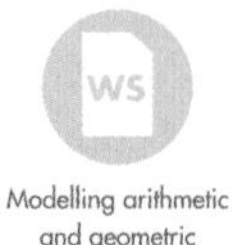

Modelling arithmetic and geometric sequences

Growth and decay – again!

Consider the growth in the value of a house that is initially valued at \$500 000 and is subject to an annual increase in value of 6.4%.

$$\text{Initial value} = \$500\,000 \quad \leftarrow \quad T_1$$
$$\text{Value after 1 year} = \$500\,000 \times 1.064 \quad \leftarrow \quad T_2$$
$$\text{Value after 2 years} = \$500\,000 \times 1.064^2 \quad \leftarrow \quad T_3$$
$$\text{Value after 3 years} = \$500\,000 \times 1.064^3 \quad \leftarrow \quad T_4, \quad \text{etc.}$$

These values form a geometric sequence with

$$T_{n+1} = T_n \times 1.064 \qquad \text{and} \qquad T_1 = 500\,000$$

Asked a question like: *At this rate how many years will it take for the value of this house to reach a value of \$1 000 000?*

We could use our ability to solve exponential equations, as covered in earlier chapters, and recognising that after x years the value will be $\$500\,000 \times 1.064^x$, simply ask a calculator to solve the equation

$$\$500\,000 \times 1.064^x = \$1\,000\,000$$

To obtain the value $x = 11.17$ (correct to 2 decimal places)

Hence the value of the house will be \$1 000 000 shortly after the end of the 11th year, i.e. early in the 12th year.

Alternatively, if we wanted to see the progressive year by year values, we could display the terms of our sequence on a calculator or spreadsheet, as shown on the next page.

ISBN 9780170390408

	A	B	C	D
1	Initial value			$500,000.00
2	Percentage increase			6.40
3	Value at end of year		1	$532,000.00
4			2	$566,048.00
5			3	$602,275.07
6			4	$640,820.68
7			5	$681,833.20
8			6	$725,470.52
9			7	$771,900.64
10			8	$821,302.28
11			9	$873,865.63
12			10	$929,793.03
13			11	$989,299.78
14			12	$1,052,614.96

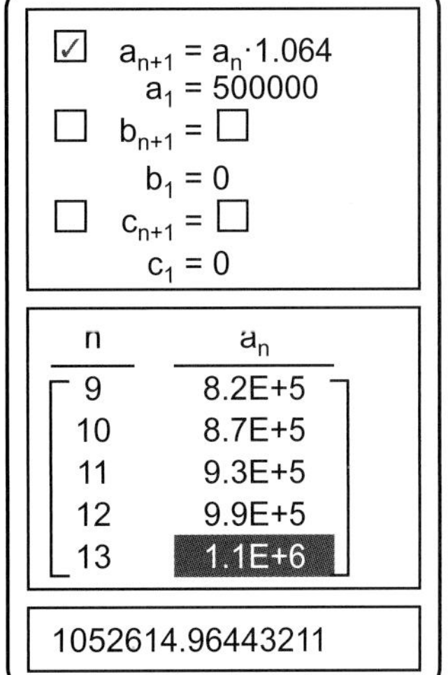

Create a spreadsheet like that shown above yourself.

As before, the value of the house will be \$1 000 000 shortly after the end of the eleventh year, i.e. early in the 12th year.

However, note carefully that in this situation, with the recursive definition

$$T_{n+1} = T_n \times 1.064 \qquad \text{and} \qquad T_1 = 500\,000$$

T_1 is the value after **zero** years. Hence we must remember that if we use the ability of a calculator to generate the terms of the sequence, according to the recursive rule given above, then the balance after n years will be given by T_{n+1}. That is, in the calculator display above, $n = 13$ gives the value at the end of 12 years.

One way to avoid this possible source of confusion would be to use the ability of some calculators to accept a sequence defined using T_0 as the 1st term.

That is, define the sequence as:

$$T_{n+1} = T_n \times 1.064 \qquad \text{and} \qquad T_0 = 500\,000,$$

as shown on the right.

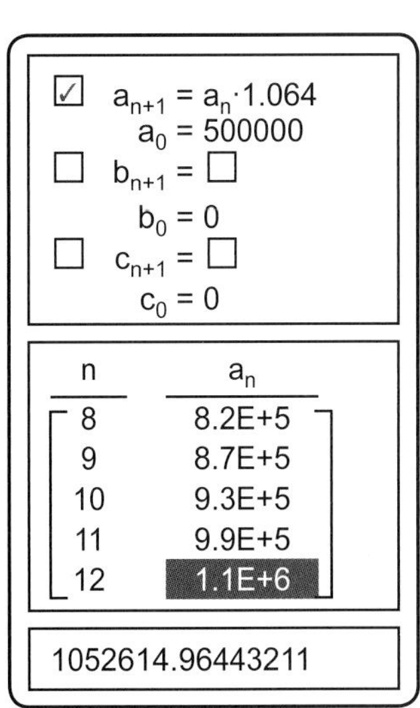

Under such a definition T_n would indeed be the value after n years.

Alternatively, we could use

$$T_{n+1} = T_n \times 1.064 \qquad \text{and} \qquad T_1 = 500\,000 \times 1.064$$

and again T_n would be the value after n years.

ISBN 9780170390408

Exercise 3C

Without using the sequence display routine available on some calculators, determine the one-hundredth term in each of the following arithmetic sequences.

1 11, 16, 21, 26, 31, 36, ...

2 –8, –5, –2, 1, 4, 7, ...

3 $T_{n+1} = T_n + 8$ with $T_1 = 23$.

4 $T_{n+1} = T_n - 2$ with $T_1 = 78$.

Without using the sequence display routine available on some calculators, determine the twenty-fifth term in each of the following geometric sequences, leaving your answers in the form $a \times b^n$.

5 5, 10, 20, 40, 80, 160, ...

6 1.5, 6, 24, 96, 384, 1536, ...

7 $T_{n+1} = 3T_n$ with $T_1 = 8$.

8 $T_{n+1} = 2T_n$ with $T_1 = 11$.

Use the ability of some calculators to display the terms of a sequence to determine the requested term in each of the following sequences.

9 $T_{n+1} = T_n + 8$ with $T_1 = 7$. Determine T_{28}.

10 $T_{n+1} = 35 - 2T_n$ with $T_1 = 5$. Determine T_{20}.

11 $T_{n+1} = 3T_n + 2$ with $T_1 = 1$. Determine T_{19}.

12 $T_{n+1} = (-1)^n T_n + 3$ with $T_1 = 6$. Determine T_{45}.

13 Julie starts a new job at a factory manufacturing automobile components. The machine she operates requires several weeks before the operator is fully accustomed to it and so her output increases each day for the first 3 weeks (15 days). On the first day she successfully completes 48 items on the machine and increases this by 3 each day after that up to and including her 15th day on the machine.

Shutterstock.com/Vasily Smirnov

Express the number of items completed on each of the first 15 days as a sequence using recursive notation.

How many items does she successfully complete on this 15th day on the machine?

14 Use the formula for the nth term of an AP with common difference d and $T_1 = a$, i.e. the formula $T_n = a + (n - 1)d$, to explain why for this AP, when we plot T_n on the y-axis and n on the x-axis, the points obtained lie on a straight line of gradient d. Find the coordinates of the point where this straight line cuts the y-axis.

15 Use the formula for the nth term of a GP with common ratio r and $T_1 = a$ to explain why for this GP, when we plot T_n on the y-axis and n on the x-axis, the points obtained fit an exponential curve. Find the equation of this curve and the coordinates of the point where it cuts the y-axis.

ISBN 9780170390408

16 Write a few sentences explaining what happens to the terms of the following arithmetic progression as $n \to \infty$.

$T_1 = a, \quad T_2 = a + d, \quad T_3 = a + 2d, \quad T_4 = a + 3d, \quad \ldots, \quad T_n = a + (n-1)d, \quad \ldots$

17 Write a few sentences explaining what happens to the terms of the following geometric progression as $n \to \infty$.

$T_1 = a, \quad T_2 = ar, \quad T_3 = ar^2, \quad T_4 = ar^3, \quad \ldots, \quad T_n = ar^{n-1}, \quad \ldots$

18 An arithmetic sequence has a first term of 8 and a common difference of 3. Determine the first four terms, the 50th term and the 100th term of the sequence.

19 An arithmetic sequence has a first term of 100 and a common difference of –3. Determine the first four terms, the 50th term and the 100th term of this sequence.

20 A geometric sequence has a first term of 11 and a common ratio of 2. Determine the first four terms, the 15th term and the 25th term of this sequence.

21 A geometric sequence has a first term of 2048 and a common ratio of 0.5. Determine the first four terms and the 16th term of this sequence.

22 Find an expression for T_n in terms of n for each of the following APs.

a 9, 15, 21, 27, 33, ...

b 7, 8.5, 10, 11.5, 13, ...

23 Find an expression for T_n in terms of n for each of the following GPs.

a 3, 6, 12, 24, 48, ...

b 100, 110, 121, 133.1, 146.41, ...

24 For the arithmetic sequence: 2, 9, 16, 23, ...

Determine **a** T_{123}

b T_{500}

c which term of the sequence is the first to exceed 1 000 000.

25 For the geometric sequence: 0.0026, 0.013, 0.065, 0.325, ...

Determine **a** T_{12}

b which term of the sequence is the first to exceed 1 000 000.

26 For the GP: 20 000 000, 15 000 000, 11 250 000, 8 437 500, ...

Determine **a** T_{12}, giving your answer to the nearest hundred

b which term of the sequence is the first less than 1.

27 The nth term of a sequence is given by $T_n = n^3$. Obtain the first four terms of this sequence and state whether the sequence is arithmetic, geometric or neither of these.

28 An arithmetic sequence has a 19th term of 61 and a 41st term of 127.

Find **a** the 20th term

b the 1st term.

ISBN 9780170390408

29 An arithmetic sequence has a 50th term of 1853 and a 70th term of 1793. Find

a the 51st term **b** the 1st term.

30 A geometric sequence has a 10th term of 98 415 and a 13th term of 2 657 205. Find

a the 14th term **b** the 1st term.

31 A geometric sequence has a 7th term of 28 672, a 9th term of 458 752 and a negative common ratio. Find

a the 10th term, **b** the 1st term.

Note for questions 32 and 33

Do each of the following two questions twice.

First using an exponential function approach, as in previous chapters,
and then by formulating a recursive formula for a sequence and then viewing the terms of the sequence on a calculator or spreadsheet.

32 \$4000 is invested into an account paying interest at 8%, compounded annually. Determine (to the nearest cent) the amount in the account at the end of ten years.

33 If a house currently valued at \$600 000 were to gain in value at 5.6% per annum, compounded annually, when would its value first exceed two million dollars?

Shutterstock.com/Artazum

Note for questions 34 and 35

The next two questions involve an initial amount being invested into an account paying interest and each year a further amount being added (question 34) or subtracted (question 35). The amounts in the account each year no longer progress geometrically but the questions can be solved using the ability of some calculators to display the terms of a sequence defined recursively.

34 \$4000 is invested into an account paying interest at 8%, compounded annually and an extra \$200 is invested after each 12 months. Thus:

Amount in account at end of 1 yr $= \$4000 \times 1.08 + \$200 \quad \leftarrow T_1$

Amount in account at end of 2 yrs $= (\$4000 \times 1.08 + \$200) \times 1.08 + \$200 \quad \leftarrow T_2$

Express T_{n+1} in terms of T_n and determine (nearest cent) the amount in the account at the end of ten years, after the \$200 for that year has been added.

35 \$4000 is invested into an account paying interest at 8%, compounded annually and \$200 is withdrawn from the account after each 12 months. Thus:

Amount in account at end of 1 yr $= \$4000 \times 1.08 - \$200 \quad \leftarrow T_1$

Amount in account at end of 2 yrs $= (\$4000 \times 1.08 - \$200) \times 1.08 - \$200 \quad \leftarrow T_2$

Express T_{n+1} in terms of T_n and determine (nearest cent) the amount in the account at the end of ten years after the \$200 for the year has been withdrawn.

ISBN 9780170390408

Miscellaneous exercise three

This miscellaneous exercise may include questions involving the work of this chapter, the work of any previous chapters, and the ideas mentioned in the Preliminary work section at the beginning of the book.

1 For each of the following, state whether the relationship between the variables x and y is linear, quadratic, exponential or reciprocal.

a $y = x^2$
b $y = 2^x$
c $y = 5x - 7$
d $y = x^2 - 3x + 4$
e $y = \frac{4}{x}$
f $y = \frac{x}{4}$
g $y = 0.5x - 12$
h $y - 6x = x^2 + 7$
i $y = (x - 5)(x + 4)$
j $xy = 7$
k $y + 8 = 2x$
l $y = 6 \times 3^x$

2 Our knowledge of the powers of 2 allows us to solve the equation $2^x = 8$ easily: $x = 3$.

This answer is also evident from the graph of $y = 2^x$ shown on the right, if we find the x value for which y, and hence 2^x, equals 8.

Use the graph to estimate solutions to the following equations:

a $2^x = 4.8$
b $2^x = 6.2$
c $2^x = 2.6$

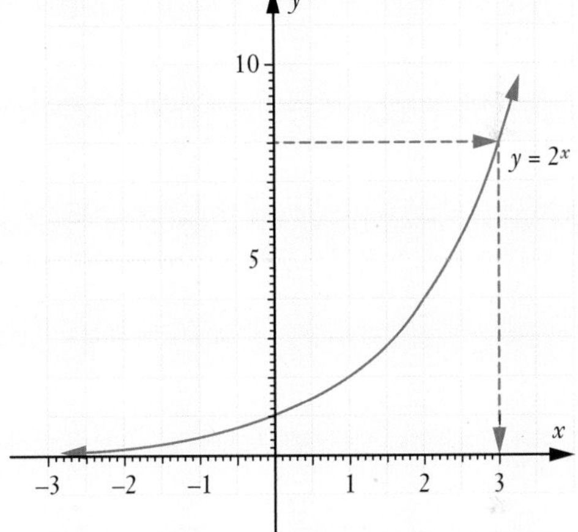

3 If the following are all written in the form 2^n, determine the value of n for each case.

a 8
b $\frac{1}{8}$
c $\frac{1}{2}$
d $\sqrt{2}$
e 1
f $\sqrt{8}$
g $\frac{1}{64}$
h $2\sqrt{2}$

4 Determine a formula for T_n, the nth term of a geometric sequence, for which $T_2 = 6$ and $T_5 = 20.25$, giving your answer in the forms

a $T_n = k \times r^{n-1}$
b $T_n = k \times r^n$

5 Find the 11th term of the geometric sequence that commences $1, \sqrt{3}, 3, \ldots$

ISBN 9780170390408

6 Determine the value of x in each of the following.

a $4^x = 64$ **b** $4^x = \frac{1}{64}$ **c** $4^x = 0.25$

d $64^{0.5} = x$ **e** $x^2 = 64$ **f** $4^8 = 4^x \times 4^{-3}$

7 Evaluate each of the following without using a calculator.

a $16^{0.5}$ **b** $16^{\frac{3}{2}}$ **c** $27^{\frac{2}{3}}$ **d** $25^{-0.5}$ **e** $\left(\frac{1}{4}\right)^{-0.5}$

8 A sequence has the recursive formula $T_{n+1} = (-1)^n T_n$, with $T_1 = 4$.

a Substitute $n = 1$ into the formula to determine T_2.

b Determine T_3 to T_5.

c Is the sequence arithmetic, geometric, or neither of these?

9 A sequence has the recursive formula $T_{n+1} = (-1)^n 2T_n$, with $T_1 = 1$.

a Substitute $n = 1$ into the formula to determine T_2.

b Determine T_3 to T_5.

c Is the sequence arithmetic, geometric or neither of these?

10 An arithmetic sequence has a first term of $5k + 3$, and a common difference of $5 - k$.

a Find in terms of k an expression for the tenth term of the sequence, simplifying your answer where possible.

b If the 20th term of this sequence is 91, find the value of the 21st term.

For numbers 11 to 19, simplify each expression without the assistance of a calculator, expressing your answers in terms of positive indices.

11 $a^4 \times a^3$ **12** $4x^2y \times 3xy^3$ **13** $(15a^3b) \div (10ab^3)$

14 $(-3a)^2 \times (2a^2b)^3$ **15** $\frac{(-3a)^2}{(2a^2b)^3}$ **16** $\frac{6a^{-1}}{(8b)^{-1}}$

17 $\frac{6a^2b^{-4}}{3a^{-3}b}$ **18** $\frac{k^7 + k^3}{k^3}$ **19** $\frac{p^5 - p^8}{p^2}$

Without the assistance of a calculator, evaluate each of the following.

20 $\frac{5^{k+2}}{5^{k-1}}$ **21** $\frac{5^{n+2} - 50}{5^n - 2}$ **22** $\frac{2^{h+3} + 8}{3 \times 2^h + 3}$

ISBN 9780170390408

4.

Series

- Arithmetic series
- A formula for S_n of an arithmetic series
- Geometric series
- A formula for S_n of a geometric series
- Infinite geometric series
- Miscellaneous exercise four

Consider the pile of cans shown below left.

The top row contains 3 cans, the second row 5 cans, the third row 7 cans and so on.

A more stable arrangement could be the arrangement shown below right.

In this case the top row contains 3 cans, the second row 4 cans, the third row 5 cans and so on.

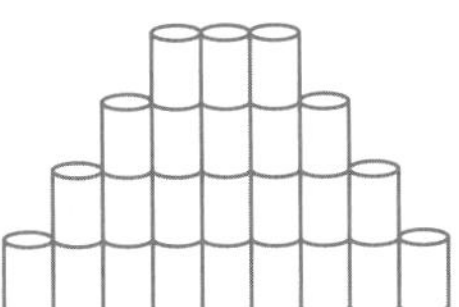

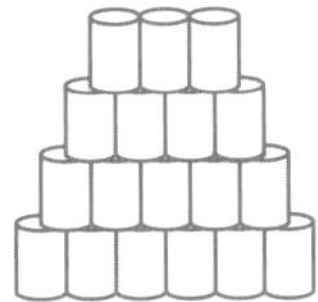

For both situations the number of cans in each row progress arithmetically. Numbering the rows from the top we have

Arithmetic sequence	Arithmetic sequence
First term 3	First term 3
Common difference 2	Common difference 1

Thus in each case we could determine the number of cans in the 20th row:

$$T_{20} = 3 + 19 \times 2 \qquad\qquad T_{20} = 3 + 19 \times 1$$
$$= 41 \qquad\qquad = 22$$

However if we were going to attempt to build these 20 row piles of cans we would also want to know how many cans each design would require altogether. I.e. we would want to evaluate

$$3 + 5 + 7 + 9 + 11 + 13 + \ldots + 41 \qquad \text{and} \qquad 3 + 4 + 5 + 6 + 7 + 8 + \ldots + 22$$

As we saw in the previous chapter, we could define each sequence recursively and use a calculator with the ability to show both the terms and partial sums of such a sequence, as shown below for the arithmetic sequence with first term 3 and common difference 2.

$a_{n+1}=a_n+2$

n + 1	a_{n+1}	Σa_{n+1}
1	3	3
2	5	8
3	7	15
4	9	24

24

Scroll down →

$a_{n+1}=a_n+2$

n + 1	a_{n+1}	Σa_{n+1}
17	35	323
18	37	360
19	39	399
20	41	440

440

When we sum the terms of a sequence we produce a **series**.

Thus 3, 5, 7, 9, 11, 13, 15, is a sequence

3 + 5 + 7 + 9 + 11 + 13 + 15 is the corresponding series

Arithmetic series

Arithmetic series

An arithmetic series has the form

$$a + (a + d) + (a + 2d) + (a + 3d) + (a + 4d) + (a + 5d) + \ldots.$$

Using S_n for the sum of the first n terms

$$\begin{aligned}
S_1 &= T_1 &&= a \\
S_2 &= T_1 + T_2 &&= a + (a + d) \\
S_3 &= T_1 + T_2 + T_3 &&= a + (a + d) + (a + 2d) \\
S_4 &= T_1 + T_2 + T_3 + T_4 &&= a + (a + d) + (a + 2d) + (a + 3d) \\
S_5 &= T_1 + T_2 + T_3 + T_4 + T_5 &&= a + (a + d) + (a + 2d) + (a + 3d) + (a + 4d)
\end{aligned}$$

$$\begin{aligned}
S_n &= T_1 + T_2 + T_3 + T_4 + T_5 + T_6 + \ldots + T_n \\
&= a + (a + d) + (a + 2d) + (a + 3d) + (a + 4d) + (a + 5d) + \ldots + a + (n - 1)d
\end{aligned}$$

EXAMPLE 1

For the arithmetic sequence 3, 7, 11, 15, 19, 23, … determine

a S_2 **b** S_5 **c** S_7.

Solution

a $S_2 = T_1 + T_2$
$= 3 + 7$
$= 10$

b $S_5 = T_1 + T_2 + T_3 + T_4 + T_5$
$= 3 + 7 + 11 + 15 + 19$
$= 55$

c $S_7 = T_1 + T_2 + T_3 + T_4 + T_5 + T_6 + T_7$
$= 3 + 7 + 11 + 15 + 19 + 23 + 27$
$= 105$

EXAMPLE 2

An arithmetic sequence has an nth term given by $T_n = 3n - 2$. Determine the first five terms of this sequence and hence determine S_1, S_2, S_3, S_4 and S_5, the first five terms of the corresponding series.

Solution

$T_n = 3n - 2 \quad \therefore$

$T_1 = 3(1) - 2 = 1$ and $S_1 = 1$

$T_2 = 3(2) - 2 = 4$ and $S_2 = 1 + 4 = 5$

$T_3 = 3(3) - 2 = 7$ and $S_3 = 1 + 4 + 7 = 12$

$T_4 = 3(4) - 2 = 10$ and $S_4 = 1 + 4 + 7 + 10 = 22$

$T_5 = 3(5) - 2 = 13$ and $S_5 = 1 + 4 + 7 + 10 + 13 = 35$

ISBN 9780170390408

Alternatively, questions like the previous two examples could be solved using a spreadsheet or using the ability of some calculators to display the terms and partial sums of sequences. However, whilst use of such technology can be very useful at times do note that for the previous two examples the 'pencil and paper' methods demonstrated were really very straightforward.

A formula for S_n of an arithmetic series

$$S_n = T_1 + T_2 + T_3 + \dots + T_{n-2} + T_{n-1} + T_n$$

Thus for the general arithmetic series with first term a and common difference d:

$$S_n = a + a+d + a+2d + \dots + a+(n-3)d + a+(n-2)d + a+(n-1)d$$

If we write these terms in the reverse order it also follows that:

$$S_n = a+(n-1)d + a+(n-2)d + a+(n-3)d + \dots + a+2d + a+d + a$$

Adding these two versions of S_n gives:

$$2S_n = \underbrace{2a+(n-1)d + 2a+(n-1)d + 2a+(n-1)d + \dots + 2a+(n-1)d + 2a+(n-1)d + 2a+(n-1)d}_{n \text{ lots of } 2a+(n-1)d}$$

$$= n[2a+(n-1)d]$$

Thus for an arithmetic progression with first term a and common difference d:

$$S_n = \frac{n}{2}[2a+(n-1)d]$$

If we use ℓ for the last term, instead of $a + (n-1)d$, this rule can be written as

$$S_n = \frac{n}{2}[a+\ell]$$

EXAMPLE 3

For the arithmetic series 10 + 17 + 24 + 31 + 38 + … determine

a S_2, **b** S_{50}.

Solution

a $S_2 = 10 + 17$

$= 27$

b Using $S_n = \frac{n}{2}[2a + (n-1)\,d]$ with $n = 50$, $a = 10$ and $d = 7$ gives

$$S_{50} = \frac{50}{2}[2(10) + 49(7)]$$

$$= 9075$$

Make sure you can obtain these same answers using

- a calculator capable of displaying the terms and sums of a sequence.
- a spreadsheet.

ISBN 9780170390408

EXAMPLE 4

For the sequence 9, 13, 17, 21, 25, 29

a Show that 157 is the 38th term. **b** Evaluate $9 + 13 + 17 + 21 + 25 + 29 + \ldots 157$.

Solution

a For an arithmetic progression with first term a and common difference d,

$$T_n = a + (n - 1)d$$

Thus for the given AP: $T_{38} = 9 + 37(4)$

$= 157$ as required.

b Using $S_n = \frac{n}{2}[a + \ell]$ with $n = 38$, $a = 9$ and $\ell = 157$

$$S_{38} = \frac{38}{2}[9 + 157]$$

$$= 3154$$

Thus $9 + 13 + 17 + 21 + 25 + 29 + \ldots 157 = 3154$

EXAMPLE 5

A fifteen-year service contract involves a company paying \$12 000 in the first year of a contract with an annual increase of \$800 every year after that for the life of the contract.

How much will the company have paid on this contract in total by the end of the fifteen years?

Solution

We require $S_{15} = \$12\,000 + \$12\,800 + \$13\,600 + \ldots + T_{15}$.

Using $S_n = \frac{n}{2}[2a + (n - 1)d]$ with $n = 15$, $a = 12\,000$ and $d = 800$ gives

$$S_{15} = \frac{15}{2}[2(12\,000) + 14(800)]$$

$$= \$264\,000$$

The company will have paid \$264000 under this contract by the end of the fifteen years.

EXAMPLE 6

A company borrows \$24000. They agree that at the end of each month the remaining balance is calculated and the company pays 1% of this remaining balance as interest payments, and then \$1000 to reduce the remaining balance. In this way the loan will be repaid in two years. How much will the loan cost the company in interest payments?

ISBN 9780170390408

Solution

End of month	1	2	3	4	5	...	24
Remaining balance	\$24 000	\$23 000	\$22 000	\$21 000	\$20 000	...	\$1000
1% of remaining balance	\$240	\$230	\$220	\$210	\$200	...	\$10

Total of interest payments $\quad \$240 + \$230 + \$220 + \$210 + \$200 + \ldots + \$10$

$$= \frac{24}{2}[\$240 + \$10]$$

$$= \$3000$$

The company pays \$3000 in interest repayments.

Note: These 'real life questions' need care. It is easy to confuse a year number with a term number and they may not always match. It is wise to list the first few terms and to think carefully how many terms are required, what is the first term, what is the common difference etc.

Exercise 4A

1 For the arithmetic sequence 8, 14, 20, 26, 32, … determine

a S_4 **b** S_5 **c** S_6

2 For the arithmetic sequence 28, 25, 22, 19, 16, … determine

a S_2 **b** S_6 **c** S_1

3 For the arithmetic sequence –6, –3, 0, 3, 6, … determine

a S_2 **b** S_5 **c** S_6

4 An arithmetic sequence has an nth term given by $T_n = 5n + 1$. Determine the first four terms of this sequence and hence determine S_1, S_2, S_3 and S_4, the first four terms of the corresponding series.

5 An arithmetic sequence is defined by the recursive rule:

$$T_{n+1} = T_n + 3, \quad T_1 = 11.$$

Determine the first four terms of this sequence and hence determine S_1, S_2, S_3 and S_4, the first four terms of the corresponding series.

6 An arithmetic sequence has an nth term given by $T_n = 25 - 3n$. Determine the first four terms of this sequence and hence determine S_1, S_2, S_3 and S_4, the first four terms of the corresponding series.

7 Determine the first 5 terms of a sequence given that the corresponding series is such that:
$S_1 = 25, S_2 = 57, S_3 = 96, S_4 = 142, S_5 = 195.$
Is the sequence arithmetic?

8 Determine the first 5 terms of a sequence given that the corresponding series is such that:
$S_1 = 1, S_2 = 5, S_3 = 14, S_4 = 30, S_5 = 55.$
Is the sequence arithmetic?

ISBN 9780170390408

9 For the arithmetic series 5 + 16 + 27 + 38 + 49 + 60... determine

a S_3 **b** S_{40}

10 For the arithmetic series 60 + 58 + 56 + 54 + 52 + 50... determine

a S_3 **b** S_{60}

11 Find the sum of the first 100 counting numbers: 1, 2, 3, 4, 5, … , 100.

12 For the sequence 16, 20, 24, 28, 32, 36, …

a Show that the 29th term is 128.

b Evaluate 16 + 20 + 24 + 28 + 32 + 36 + … + 128.

13 For the sequence 9, 26, 43, 60, 77, 94, …

a Show that the 41st term is 689.

b Evaluate 9 + 26 + 43 + 60 + 77 + 94 + … + 689.

14 Part of a cyclist's training program involves her in riding 20 km on the first day of the month, 22 km on the second day, 24 on the third and so on, the distances increasing in arithmetic progression.

How far will she cycle on the 30th day of the month?

How far will she cycle in total during these 30 days?

Shutterstock.com/Dimitry Morgan

15 A farmer plants trees in a corner of one of his paddocks. He plants them in rows with the first row containing 5 trees, the second containing 7 trees, the third containing 9 trees and so on. How many trees will he need to plant if he is to plant 15 rows?

16 A twelve month equipment hire contract involves a company paying $4000 at the end of the first month, $3750 at the end of the second month, and so on, with each monthly payment decreasing by $250 after that, for the life of the contract. How much will the company have paid in total on this contract by the time they make the last payment at the end of the twelfth month?

17 Jack is offered two jobs, one with company A and the other with company B.

Company A offers him $65 000 in the first year, increasing by $2500 in each subsequent year.

Company B offers him $68 000 in the first year, increasing by $1200 in each subsequent year.

How much would he receive from each company if he were to work for them for ten years?

18 A company borrows $36 000. They agree that at the end of each month the remaining balance is calculated and the company pays 2% of this remaining balance, as interest payments, and then pays $2000 to reduce the remaining balance. In this way they will repay the loan in eighteen months. How much will the loan cost the company in interest payments?

ISBN 9780170390408

Geometric series

Geometric series

Geometric series 2

Mixed sequences and series

EXAMPLE 7

A geometric sequence has an nth term given by $T_n = 3(2)^n$.

Determine the first five terms of this sequence and hence determine S_1, S_2, S_3, S_4, and S_5, the first five terms of the corresponding geometric series.

Solution

$$T_n = 3(2)^n \quad \therefore \quad \begin{array}{lll} T_1 = 3(2)^1 = 6 & \text{and} & S_1 = 6 \\ T_2 = 3(2)^2 = 12 & \text{and} & S_2 = 6 + 12 = 18 \\ T_3 = 3(2)^3 = 24 & \text{and} & S_3 = 6 + 12 + 24 = 42 \\ T_4 = 3(2)^4 = 48 & \text{and} & S_4 = 6 + 12 + 24 + 48 = 90 \\ T_5 = 3(2)^5 = 96 & \text{and} & S_5 = 6 + 12 + 24 + 48 + 96 = 186 \end{array}$$

A formula for S_n of a geometric series

For the general geometric series with first term a and common ratio r:

$$S_n = a + ar + ar^2 + ar^3 + \ldots + ar^{n-2} + ar^{n-1} \qquad [1]$$

Multiply by r:

$$rS_n = ar + ar^2 + ar^3 + ar^4 + \ldots + ar^{n-1} + ar^n \qquad [2]$$

Equation [1] – equation [2]:

$$S_n - rS_n = a + 0 + 0 + 0 + \ldots\ldots\ldots\ldots + 0 - ar^n$$

$$S_n(1 - r) = a - ar^n$$

For the geometric sequence with first term a and common ratio r

$$S_n = \frac{a(1-r^n)}{1-r} \quad r \neq 1$$

If r is not between –1 and 1 this formula is easier to use in the form

$$S_n = \frac{a(r^n-1)}{r-1} \quad r \neq 1$$

EXAMPLE 8

Evaluate S_{10} for the series 512 + 768 + 1152 + 1728 + …

Solution

The given sequence of terms are geometric with first term 512 and common ratio 1.5.

Using $S_n = \dfrac{a(r^n-1)}{r-1}$

$$S_{10} = \frac{512(1.5^{10}-1)}{1.5-1}$$

$$= 58\,025$$

ISBN 9780170390408

EXAMPLE 9

A geometric progression has a first term of 40 and a common ratio of 0.75. Find S_{12}, the sum of the first twelve terms, giving your answer correct to two decimal places.

Solution

In this case r is between –1 and 1 so we use $S_n = \dfrac{a(1-r^n)}{1-r}$

$$S_{12} = \frac{40(1-0.75^{12})}{1-0.75}$$

$$= 154.93 \text{ correct to 2 decimal places}$$

EXAMPLE 10

What is the least number of terms required for the geometric series

$$10 + 15 + 22.5 + 33.75 + \ldots$$

to have a sum that exceeds 1 000 000?

Solution

In this case $r = 1.5$ and $a = 10$. Use $S_n = \dfrac{a(r^n-1)}{r-1}$ and let $S_n = 1\,000\,000$

i.e. $$1\,000\,000 = \frac{10(1.5^n - 1)}{1.5-1}$$

$$50\,000 = 1.5^n - 1$$

By trial and adjustment n = bigger than 26 but less than 27.

However n must be a positive integer. Thus at least 27 terms are needed for the given sum to exceed 1 000 000. (Alternatively $50\,000 = 1.5^n - 1$ can be solved using a calculator with a 'solve facility'.)

EXAMPLE 11

On 1 January 2014, Mrs Smith starts a savings plan by investing \$500 into an account that guarantees 5% interest per annum provided she commits to invest a further \$500 on 1 January every year. The plan will finish on 31 December 2023. How much should this savings plan be worth on 31 December 2023 (to the nearest 10 cents)?

Solution

Date	Value of investment
1 Jan 2014	\$500
1 Jan 2015	$\$500 + \$500 \times 1.05$
1 Jan 2016	$\$500 + \$500 \times 1.05 + \$500 \times 1.05^2$
1 Jan 2017	$\$500 + \$500 \times 1.05 + \$500 \times 1.05^2 + \$500 \times 1.05^3$
	⋮
1 Jan 2023	$\$500 + \$500 \times 1.05 + \$500 \times 1.05^2 + \ldots + \$500 \times 1.05^9$
31 Dec 2023	$\$500 \times 1.05 + \$500 \times 1.05^2 + \ldots + \$500 \times 1.05^9 + \$500 \times 1.05^{10}$

ISBN 9780170390408

Thus the final value is S_{10} for a geometric series with 1st term \$500 × 1.05 and $r = 1.05$.

$$S_{10} = \frac{\$525(1.05^{10} - 1)}{1.05 - 1}$$
$$= \$6603.40 \text{ to the nearest 10 cents.}$$

Alternatively:

- The previous example could be solved using a recursive definition, as in the previous chapter. With $T_1 = 500$ and $T_{n+1} = T_n \times 1.05 + 500$, the required answer would be given by $T_{11} - 500$. (Or with $T_0 = 500$ and $T_{n+1} = T_n \times 1.05 + 500$, and the required answer given by $T_{10} - 500$.)
- Some calculators have built-in financial programs that will calculate values of accounts in situations like that of the last example. Whilst you are encouraged to explore the capability of your calculator in this regard make sure that if required you can calculate values using geometric series and recursively defined sequences.

Exercise 4B

1 A geometric sequence has an nth term given by $T_n = 2(3)^n$. Determine the first four terms of this sequence and hence determine S_1, S_2, S_3, and S_4, the first four terms of the corresponding series.

2 A geometric sequence is defined by the recursive rule $T_{n+1} = 1.5T_n,\ T_2 = 24$.
Determine the first four terms of this sequence and hence determine S_1, S_2, S_3 and S_4, the first four terms of the corresponding geometric series.

3 Determine the first 5 terms of a sequence given that the corresponding series is such that:
$S_1 = 1, S_2 = 2, S_3 = 4, S_4 = 7, S_5 = 12$.
Is the sequence geometric?

4 Determine the first 5 terms of a sequence given that the corresponding series is such that:
$S_1 = 8, S_2 = 32, S_3 = 104, S_4 = 320, S_5 = 968$.
Is the sequence geometric?

5 Evaluate S_{15} for the series $1 + 2 + 4 + 8 + 16 + \ldots$.

6 Evaluate S_{11} for the series $20480 + 10240 + 5120 + 2560 + 1280 + \ldots$.

7 A geometric sequence has a first term of 256 and a common ratio of 2.5. Find S_9, the sum of the first nine terms.

8 A geometric sequence has a first term of 62 500 and a common ratio of 0.4. Find S_9, the sum of the first nine terms.

9 In the spreadsheet display shown on the right, column B shows the first 7 terms of a geometric sequence. If S_n represents the sum of the first n terms determine S_6, S_7 and S_8.

	A	B
1	Term	Value
2	1	2.25
3	2	9
4	3	36
5	4	144
6	5	576
7	6	2304
8	7	9216

ISBN 9780170390408

10 What is the least number of terms required for the geometric series $5 + 10 + 20 + 40 + \ldots$ to have a sum that exceeds 5 000 000?

11 What is the least number of terms required for the geometric series $28 + 42 + 63 + 94.5 + \ldots$ to have a sum that exceeds 1 000 000?

12 A GP has a third term of 24 and a fourth term of 96. Find T_{10} and S_{10}.

13 A sports star negotiates a contract with a sports equipment company that ensures he will be paid \$50 000 in the first year, \$57 500 in the second year, \$66 125 in the third year and so on, the annual amounts continuing in this geometric progression. Calculate the total amount the star will receive in the 10 years this contract is for. (Give your answer to the nearest \$1000.)

Shutterstock.com/Tania Thompson

14 In its first year of operation a mine yielded 5000 tonnes of a particular mineral. In the second year the yield was approximately 110% of the first year's figures. In each subsequent year the yield continued to be approximately 110% of the previous year's yield.

What tonnage of the mineral did the mine yield in

a its second year of operation?

b its third year of operation?

c its fourth year of operation?

What total tonnage of the mineral did the mine yield in its first twelve years of operation?

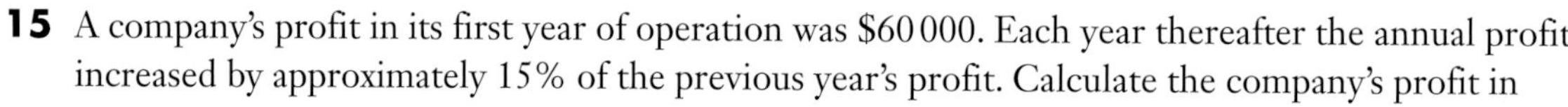

15 A company's profit in its first year of operation was \$60 000. Each year thereafter the annual profit increased by approximately 15% of the previous year's profit. Calculate the company's profit in

a the second year **b** the third year **c** the tenth year.

Find the total profit the company makes in its first ten years of operation.

16 \$1200 is deposited on 1 January 2014 to open an account. On each subsequent anniversary of this date a further \$1200 is deposited into the account. The account earns interest at the rate of 10% per annum, compounded annually.

Copy and complete the 1/1/18 line in the following table for this situation.

	Value of deposit made on					
Date	**1/1/14**	**1/1/15**	**1/1/16**	**1/1/17**	**1/1/18**	**Total value**
1/1/14	\$1200					\$1200
1/1/15	$\$1200 \times 1.1$	\$1200				\$2520
1/1/16	$\$1200 \times 1.1^2$	$\$1200 \times 1.1$	\$1200			\$3972
1/1/17	$\$1200 \times 1.1^3$	$\$1200 \times 1.1^2$	$\$1200 \times 1.1$	\$1200		\$5569.20
1/1/18						

How much will be in the account immediately following the deposit of \$1200 made on 1 January 2029, to the nearest dollar?

ISBN 9780170390408

17 A person deposits \$1000 on 1 January 2015 and a further \$1000 on 1 January of each subsequent year up to and including 2024. The investment gains interest at 7% per annum compounded annually. How much will be in the account when it is closed on 31 December 2024? (Give your answer to the nearest dollar.)

18 As a new employee becomes more accustomed to the machine she is operating, the number of units she produces per day increases. Her daily production approximates closely to a geometric progression for the first fifteen days with 2500 produced on the first day, 2550 on day two, 2601 on day three, etc. The employee maintains the fifteenth day's output thereafter.

- **a** How many units does she produce on day 4?
- **b** How many units does she produce on day 15?
- **c** How many units does she produce on day 16?
- **d** Find the total number of units this employee produces in her first fifteen days working with the company. (To the nearest hundred.)
- **e** Find the total number of units this employee produces in her first forty days working with the company. (To the nearest hundred.)

19 8000 tonnes of a particular mineral were mined in each of the first three years of a mine's operation. In the fourth year the quantity mined was 90% of the third year's output, the fifth year was 90% of the fourth year and so on. The mine was closed at the end of the first year in which the amount mined fell below 1900 tonnes.

- **a** For how many years did the mine remain open?
- **b** What total tonnage of the mineral was mined from the mine?

20 After the birth of their son, Mr and Mrs Jacques decide to open an account for him commencing when he reaches the age of 1. They wish to invest the same fixed amount on each birthday from 1 to 21 such that, immediately following the 21st payment, the account would hold \$50 000. The account earns interest at 9.5% per annum, compounded annually.

Suppose the fixed amount they wish to invest is \$*P*. The table below shows how the value of the account grows during the first four years.

	Value of deposit made on			
Birthday	**1st birthday**	**2nd birthday**	**3rd birthday**	**4th birthday**
1	$\$P$			
2	$\$P \times 1.095$	$\$P$		
3	$\$P \times 1.095^2$	$\$P \times 1.095$	$\$P$	
4	$\$P \times 1.095^3$	$\$P \times 1.095^2$	$\$P \times 1.095$	$\$P$

Thus immediately after the 4th birthday payment the account is worth

$$\$P + \$P \times 1.095 + \$P \times 1.095^2 + \$P \times 1.095^3$$

- **a** What is the first term, common ratio and number of terms for the similar expression for the value of the account immediately after the 21st payment has been made?
- **b** Determine the value of *P* correct to 1 decimal place.

Infinite geometric series

Each of the following tables involve geometric progressions.

Copy and complete the six tables.

1 $a = 64$
$r = 5$

n	T_n	S_n
1	64	64
2	320	384
3	1600	1984
4	8000	9984
5		
6		
7		

2 $a = 0.4$
$r = 4$

n	T_n	S_n
1	0.4	0.4
2	1.6	2.0
3	6.4	8.4
4	25.6	34.0
5		
6		
7		

3 $a = 100$
$r = 1.8$

n	T_n	S_n
1	100	100
2	180	280
3	324	604
4	583.2	1187.2
5		
6		
7		

4 $a = 64$
$r = 0.2$

n	T_n	S_n
1	64	64
2	12.8	76.8
3	2.56	79.36
4	0.512	79.872
5		
6		
7		

5 $a = 10$
$r = 0.5$

n	T_n	S_n
1	10	10
2	5	15
3	2.5	17.5
4	1.25	18.75
5		
6		
7		

6 $a = 90$
$r = 0.4$

n	T_n	S_n
1	90	90
2	36	126
3	14.4	140.4
4	5.76	146.16
5		
6		
7		

Reading down the S_n columns in your completed tables you should notice that in each table the numbers are getting bigger and bigger as we go down the column. However notice that whilst in the first three tables these numbers get very big, in the last three tables the numbers are increasing but by a smaller and smaller amount each time.

ISBN 9780170390408

Indeed it seems that:

the numbers in the S_n column for table 4 are heading towards 80,
the numbers in the S_n column for table 5 are heading towards 20,
the numbers in the S_n column for table 6 are heading towards 150.

This should be no surprise if we consider the formula: $S_n = \dfrac{a(1-r^n)}{1-r}$.

For tables 4, 5 and 6 the common ratio is between –1 and 1 and all such numbers get very small when raised to a large power. Thus in our formula r^n will get smaller and smaller as n gets bigger and bigger

and so, as n gets large $\quad S_n \approx \dfrac{a}{1-r}$.

For table 4, $a = 64$ and $r = 0.2$. Thus as n gets large

$$S_n \approx \frac{64}{1-0.2}$$
$$= 80, \text{ as we found.}$$

For table 5, $a = 10$ and $r = 0.5$. Thus as n gets large

$$S_n \approx \frac{10}{1-0.5}$$
$$= 20, \text{ as we found.}$$

For table 6, $a = 90$ and $r = 0.4$. Thus as n gets large

$$S_n \approx \frac{90}{1-0.4}$$
$$= 150, \text{ as we found.}$$

We call $\dfrac{a}{1-r}$ the formula for the *sum to infinity* of a GP with first term a and common ratio r.

This is the value that S_n gets closer and closer to as n gets bigger and bigger.

Remember: this concept only makes sense for geometric progressions for which

$$-1 < r < 1.$$

Using the symbol '∞' to represent infinity:

> For a geometric series $a + ar + ar^2 + ar^3 + \ldots$ with $-1 < r < 1$, $S_\infty = \dfrac{a}{1-r}$.

EXAMPLE 12

Determine the sum to infinity of a geometric progression with first term 36 and common ratio 0.25.

Solution

$$S_\infty = \frac{a}{1-r}$$
$$= \frac{36}{1-0.25}$$
$$= 48$$

The sum to infinity for the given series is 48.

The table below shows T_n and S_n for $n = 1$ to 10 for the geometric progression of the previous example, i.e. for the GP with first term 36 and common ratio 0.25 for which we found $S_\infty = 48$.

n	T_n	S_n
1	36	36
2	9	45
3	2.25	47.25
4	0.5625	47.8125
5	0.140 625	47.953 125
6	0.035 156 25	47.988 281 25
7	0.008 789 062 5	47.997 070 312 5
8	0.002 197 265 625	47.999 267 578 125
9	0.000 549 316 406 25	47.999 816 894 531 25
10	0.000 137 329 101 562 5	47.999 954 223 632 812 5

EXAMPLE 13

For each of the following geometric series determine whether S_∞ exists and, if it does, determine its value.

a $120 + 90 + 67.5 + \ldots$

b $64 + 96 + 144 + \ldots$

Solution

a

$$r = \frac{90}{120} = 0.75$$

$$S_\infty = \frac{120}{1-0.75} = 480$$

b

$$r = \frac{96}{64} = 1.5$$

r is not between -1 and 1.

S_∞ does not exist.

EXAMPLE 14

A patient's body absorbs a certain drug in such a way that whatever is in the body at a particular time, 50% remains in the body 24 hours later. Every 24 hours for the rest of his life the patient has to give himself an injection containing 20 mg of the drug.

In the long term, how many mg of the drug will be in the patient's body

a immediately after each injection?

b immediately prior to each injection?

Number of mg in patient's body immediately before and immediately after injection		
	Immediately before	**Immediately after**
1st injection	0	20
2nd injection	$0.5(20) = 10$	$10 + 20$
3rd injection	$0.5(10 + 20) = 5 + 10$	$5 + 10 + 20$
4th injection	$2.5 + 5 + 10$	$2.5 + 5 + 10 + 20$

ISBN 9780170390408

Solution

Thus in the long term the amount in the patient's body

a immediately after each injection $= 20 + 10 + 5 + 2.5 + \ldots$

$$= \frac{20}{1-0.5}$$

$$= 40 \text{ mg}$$

b immediately before each injection $= 10 + 5 + 2.5 + \ldots$

$$= \frac{10}{1-0.5}$$

$$= 20 \text{ mg}$$

Exercise 4C

1 The table below shows T_n and S_n, from $n = 1$ to $n = 8$, for 3 GPs, A, B and C.

For each of the progressions

a determine the common ratio,

b state whether S_∞ exists and, if it does, state its value.

	Geometric progression A		Geometric progression B		Geometric progression C	
n	T_n	S_n	T_n	S_n	T_n	S_n
1	24	24	8	8	35	35
2	9.6	33.6	12	20	10.5	45.5
3	3.84	37.44	18	38	3.15	48.65
4	1.536	38.976	27	65	0.945	49.595
5	0.6144	39.5904	40.5	105.5	0.2835	49.8785
6	0.245 76	39.836 16	60.75	166.25	0.085 05	49.963 55
7	0.098 304	39.934 464	91.125	257.375	0.025 515	49.989 065
8	0.039 3216	39.973 7856	136.6875	394.0625	0.007 6545	49.996 7195

2 For each of the following geometric series, determine whether S_∞ exists and, if it does, determine its value.

a $100 + 50 + 25 + \ldots$

b $100 + 75 + 56.25 + \ldots$

c $100 + 110 + 121 + \ldots$

d $90 + 72 + 57.6 + \ldots$

e $56 + 70 + 87.5 + \ldots$

f $90 - 72 + 57.6 - \ldots$

g $0.6 + 0.2 + 0.0\overline{6} + \ldots$

h $2304 - 288 + 36 - \ldots$

3 A geometric series with a first term of 48 has a sum to infinity of 120.

Determine the common ratio of this series.

4 A geometric series with a common ratio of 0.45 has a sum to infinity of 120.

Determine the first term of this series.

ISBN 9780170390408

5 A patient's body absorbs a certain drug in such a way that whatever is in the body at a particular time, 40% remains in the body 24 hours later.

Every 24 hours for the rest of his life the patient has to give himself an injection containing 15 mg of the drug.

Copy and complete the following table:

Number of mg in patient's body immediately before and after injection		
	Immediately before	**Immediately after**
1st injection	0	15
2nd injection		
3rd injection		
4th injection		
5th injection		

In the long term how many mg of the drug will be in the patient's body

a immediately after each injection?

b immediately prior to each injection?

6 An athlete is taking part in a test to assess his endurance and fitness.

The athlete manages to complete a particular exercise 50 times in the first minute, 40 times in the second minute and 32 times in the third minute without stopping for a rest. If the athlete were to continue this activity, and if the number of times he completed the particular exercise in each minute continued the geometric progression of the first three minutes, theoretically what is the greatest number of times he could complete the exercise without stopping for a rest? Discuss the reality of this theoretical greatest number.

7 When a particular ball is dropped onto a horizontal surface the height it reaches on its first bounce is 60% of the height of the previous bounce. Subsequent bounce heights are 60% of the height of the previous bounce.

If the ball is dropped from a height of 2 metres, onto a horizontal surface, find

a the height reached on the first bounce,

b the height reached on the sixth bounce,

c the total vertical distance the ball travels until the bouncing ceases.

8 When a particular ball is dropped onto a horizontal surface the height it reaches on its first bounce is 40% of the height of the previous bounce. Subsequent bounce heights are 40% of the height of the previous bounce. If the ball is dropped from a height of 5 metres, onto a horizontal surface, find the total vertical distance the ball travels before coming to rest (to the nearest centimetre).

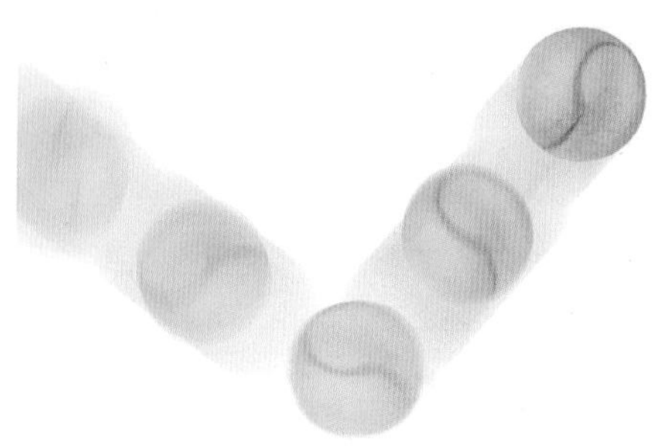

Shutterstock.com/Wichai Wonglong Jaihan

ISBN 9780170390408

Miscellaneous exercise four

This miscellaneous exercise may include questions involving the work of this chapter, the work of any previous chapters, and the ideas mentioned in the Preliminary work section at the beginning of the book.

1 Express each of the following as a power of 2.

a 64 **b** 256 **c** $2^3 \times 2 \times 2 \times 2 \times 2$

d $2 \times 2 \times 2 \times 2 \times 2 \div 2^3$ **e** $2^6 \times 2^4$ **f** $2^6 \div 2^4$

g $4 \times 8 \times 16 \times 32$ **h** 1 **i** $6^2 - 2^2$

Evaluate each of the following without the use of a calculator.

2 2^{-1} **3** $4^8 \div 4^6$ **4** $\left(\frac{3}{2}\right)^2$

5 18^0 **6** $(4^{0.5})^6$ **7** $5^6 \times 5^{-8}$

8 $\dfrac{3^7 \times 27^2}{3^{14}}$ **9** $\dfrac{5^8 \div 5^4}{125}$ **10** $\dfrac{7^{10} \div 7^2}{49^2 \times 7^5}$

11 Given that $2^n \div 2^m = 2^{n-m}$ explain why it then makes sense for 2^0 to be equal to 1.

12 Copy and complete the following table.

	$T_1, T_2, T_3, T_4, T_5, \ldots$	**Recursively defined**
a		$T_n = T_{n-1} + 5, T_1 = 17$
b		$T_n + 1 = T_n - 7, T_1 = 100$
c		$T_n = 3T_{n-1}, T_1 = 5$
d	6, 10, 14, 18, 22, …	
e	2, 6, 18, 54, 162, …	
f	17, 9, 1, –7, –15, …	

13 Each of the two graphs shown below show the terms of a sequence. One sequence involves a recursive rule of the form $T_{n+1} = T_n + a$ and the other involves a recursive rule of the form $T_{n+1} = kT_n$ where a and k are constants.

Determine the value of a and k and hence determine T_{20} for each sequence.

a

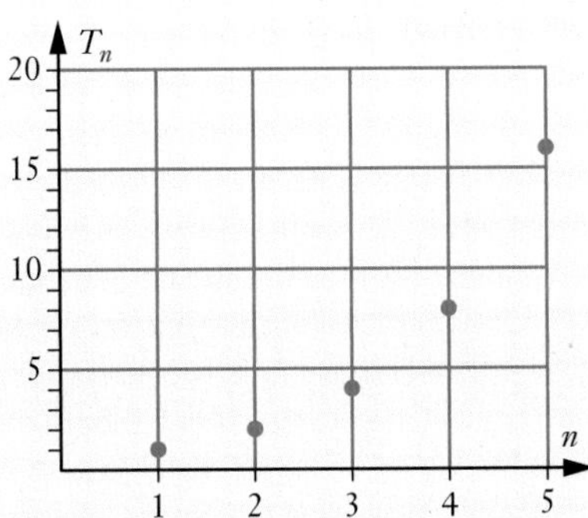

b

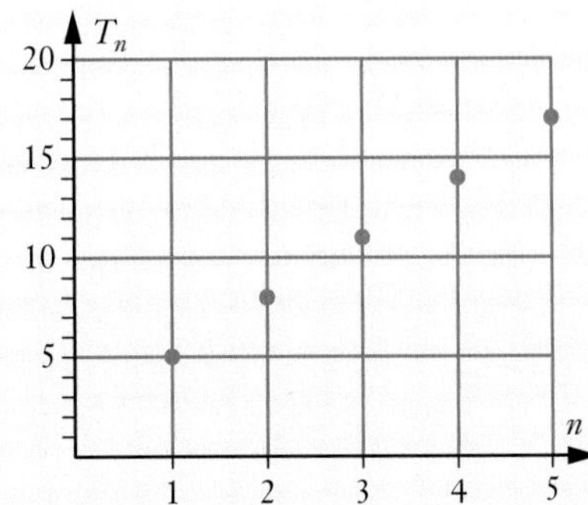

ISBN 9780170390408

14 a Without the assistance of a calculator, evaluate ${}^{6}C_4$.

b Show that ${}^{n}C_1 = n$ and ${}^{n}C_2 = \dfrac{n(n-1)}{2}$.

15 Find the value of x and define each sequence recursively if the three terms

$$8, x, 50$$

are, in that order, the first three terms of

a an arithmetic progression **b** a geometric progression.

16 The resale value of a particular item of machinery, t years after purchase, is expected to be $\$V$ where $V \approx 250\,000(0.82)^t$.

Use your calculator to view the graph of V for $0 \le t \le 15$.

Using your graph, or by some other method, determine in how many years time the value will be

a 50% of its current (i.e. $t = 0$) value **b** 25% of its current (i.e. $t = 0$) value.

17 Rosalyn does 30 minutes of dance practice each day. As the national championships approach Rosalyn decides to increase the amount she does by 3 minutes each day for the 20 days prior to the championships i.e. 20 days prior she will practise for 33 minutes, 19 days prior she will practise for 36 minutes etc.

Shutterstock.com/Alexander Y

Express the sequence 30, 33, 36, 39, 42, using recursive notation.

Under this scheme for how long will Rosalyn practise 1 day prior to the championships?

For how long does Rosalyn practise in total during these 20 days prior to the championships?

18 A competition advertises that the 1st prize is 1 million dollars. Whilst the winner of this prize will indeed receive \$1 million they will not receive it all at once. The prize conditions state that the winner will receive \$50 000 immediately the win is announced, followed by \$50 000 each year thereafter, on the anniversary of the first payment, for 19 further payments.

Compare the value of the following accounts after twenty years:

A: An account opened with a 1 million dollar investment, earning interest at 6% per annum compounded annually and left untouched for the 20 years.

B: An account opened with a \$50 000 payment followed by a further \$50 000 invested each year thereafter for 19 further payments, the account earning interest at 6% per annum compounded annually and, other than the regular annual injections of \$50 000, left untouched for 20 years.

How much do the organisers have to have available 'now' (i.e. at the time the winner is announced), rounded up to the next dollar, in order to meet their financial commitments to the winner, if they are to pay the initial \$50 000 from this 'available now' fund, and invest the rest in an account also paying a constant 6% interest per annum compounded annually, with the aim of being able to pay the nineteen annual amounts from this account, with the account balance reduced to zero at the end of this time?

ISBN 9780170390408

5.

Rates of change

- The gradient of a curve
- Calculating the gradient at a point on a curve
- General statement of this 'limiting chord' process
- Notation
- Finding the equation of a tangent at a point on $y = ax^n$
- Differentiating $f(x) \pm g(x)$
- Differentiating more general power functions
- Miscellaneous exercise five

Situation One

Shutterstock.com/Niv Koren

Research scientists are testing a new, lightweight alloy for its possible use in car, train and aeroplane bodies. In one test a vehicle made of the alloy is propelled along a straight horizontal railway in such a way that the distance, y metres, that the vehicle has travelled t seconds after it started is given by

$$y = t^3.$$

Unfortunately the vehicle starts to break up 8 seconds after it started.

The scientists knew that at that instant the vehicle was 512 metres from its starting point ($512 = 8^3$). For those 8 seconds the average speed was given by:

$$\frac{\text{distance travelled in the 8 seconds}}{\text{time taken}} = \frac{512 \text{ m}}{8 \text{ s}}$$

$$= 64 \text{ m/s} = (230.4 \text{ km/h}).$$

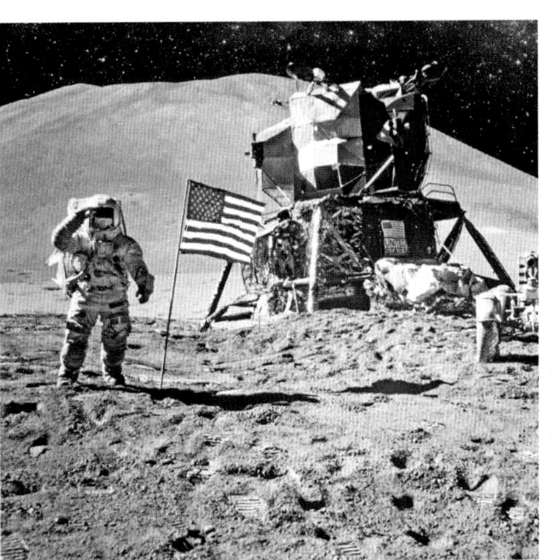
Shutterstock.com/Nasa/Castleski

However the scientists want to know the speed the vehicle was travelling at the instant that it started to break up.

Try to determine the speed of the vehicle at this instant (i.e. at $t = 8$).

Situation Two

It is the year 2035 and plans are well advanced for the building of a space station on the Moon. The space station will be pressurised and will act as a lunar laboratory and repair depot for space vehicles servicing the various telecommunication and surveillance satellites. The space station will obtain its power from thousands of solar tiles on the roof.

Health and Safety experts are concerned that tiles dislodged from the roof could fall on astronauts working outside the station and damage their space suits. They want tests to be carried out on Earth to ensure that the space suits are strong enough to withstand the impact.

For such tests to be carried out, the speed any dislodged tiles will have at the instant they reach the Moon's surface is required. It is known that the tiles will fall from a height of 20 metres. Due to the Moon's gravitational pull any tile will have fallen y metres, t seconds after it is dislodged, where

$$y = 0.8t^2.$$

- Calculate the value of t at the instant a dislodged tile hits the surface of the Moon. (Take the time the tile is dislodged as $t = 0$.)
- Calculate the average speed of the tile during its fall.
- Calculate the speed of the tile at the instant it strikes the surface of the Moon.

ISBN 9780170390408

The two situations on the previous page each involved finding the speed of an object at an instant, i.e. the **instantaneous** speed. Perhaps with the first situation, having found the average speed for the motion from $t = 0$ to $t = 8$ you then considered the average speed for the motion from $t = 7$ to $t = 8$, and then perhaps you considered … etc.

Speed is the rate at which an object changes its position with respect to time. The two situations required us to find the rate at which the distance variable (y) was changing with respect to the time variable (t), at a particular instant.

Graphically, the rate of change of one variable, y, with respect to another variable, x, is the gradient of the graph of the relationship. Therefore if either of the situations on the previous page had involved linear functions it would have been an easy matter to determine the gradient by comparison with the form $y = mx + c$, or in the situations given, $y = mt + c$. However neither $y = t^3$ nor $y = 0.8t^2$ are of this form so, as we know, neither have straight line graphs. Therefore, if we wish to pursue this gradient idea, we first need to think about what we might mean by *the gradient of a curve*.

The gradient of a curve

In the linear relationship shown graphed on the right, each unit increase in x sees a 3 unit increase in y.

The straight line has a *constant gradient*, or *slope*, of 3.

We could say that the *rate of change of y with respect to x* is 3.

If the two variables are not linearly related the gradient, or slope, is not constant. We must then refer to the gradient at a particular point.

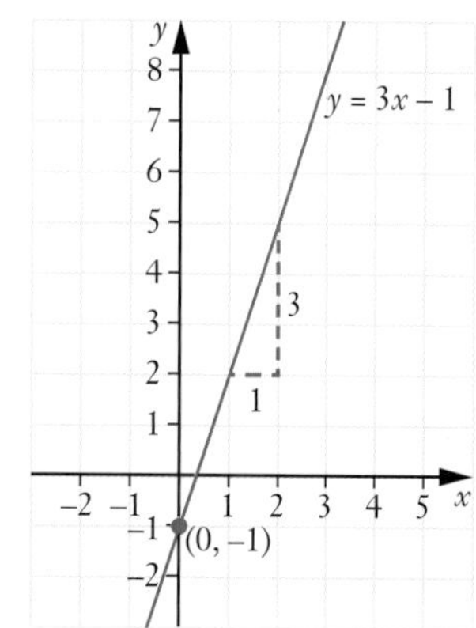

> We define the **gradient** at some point P on the curve $y = f(x)$ to be the **gradient** of the **tangent** to the curve at the point P.

The tangent at P is the line that 'just touches' the curve at that point (except if P is a point of inflection as we will see on the next page).

If we can determine the gradient of this tangent we know the gradient of the curve at the point P and hence the rate of change of y with respect to x at $x = a$.

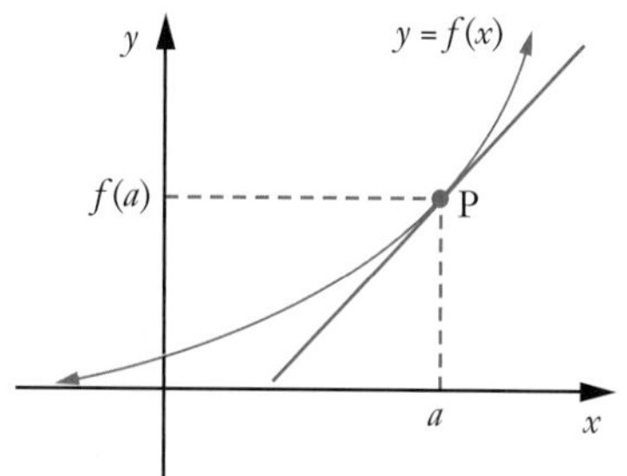

Consider the graph shown on the right and in particular consider the positive or negative nature of the gradient of this curve at the points P, Q, R, S and T.

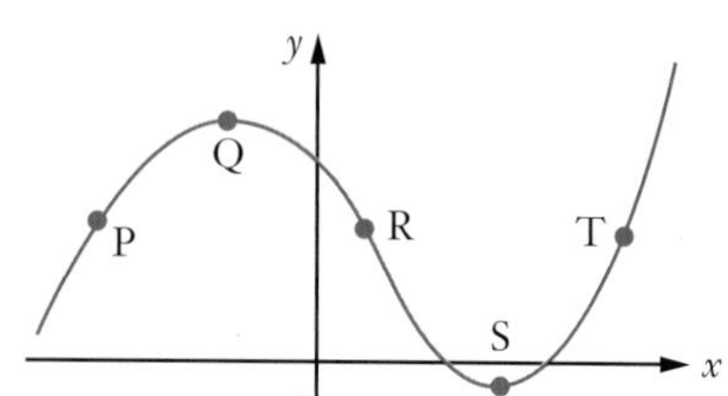

ISBN 9780170390408

The diagram on the right now has the tangents to the curve at points P, Q, R, S and T drawn.

From this we can notice the following.

At points P and T the gradient is positive.
(Uphill for increasing x.)
At point R the gradient is negative.
(Downhill for increasing x.)
At points Q and S the gradient is zero. (The tangent is horizontal.)

Between points P and Q we say the function is *increasing* (a positive gradient).
Between points Q and S we say the function is *decreasing* (a negative gradient).
Between points S and T the function is *increasing* (a positive gradient).

The *Preliminary work* section at the beginning of this book reminded us of some other vocabulary used to describe some key features of the graphs of functions. The following dot points expand on some of this vocabulary, again referring to the graph shown above.

- In the graph the point Q is a **maximum turning point** and point S is a **minimum turning point**. The gradient of the curve is zero at these two points. (The tangents at these points are horizontal.)
- We can also refer to point Q as a **local maximum point** (sometimes referred to as a *relative* maximum point). There may be points on the graph that are higher than Q but *in the locality* of point Q it is the highest point.
- Looking at the section of the graph displayed the highest point overall is at the right hand end. This would be the **global maximum** for the section shown.
- Similarly point S is a **local minimum point** (or *relative* minimum point) and, for the section of graph shown, S is also the **global minimum**.
- From the extreme left of the display to point R the curve is **concave down**.
- From the point R to the extreme right of the display the curve is **concave up**.
- Point R, where the concavity changes, is a **point of inflection**. (Notice that at this point the tangent line actually cuts the curve.)
- If the graph were to continue its upward path as x increases then as x gets large positively (we say 'as x **tends to infinity**' and we write '$x \to \infty$') then y also gets large positively.
  We write: As $x \to +\infty$ then $y \to +\infty$.
  And similarly: As $x \to -\infty$ then $y \to -\infty$.
- Local maximum and local minimum points are sometimes referred to as **turning points**. At all such points the gradient is zero.
- Maximum points, minimum points and points of horizontal inflection are sometimes referred to as **stationary points**.

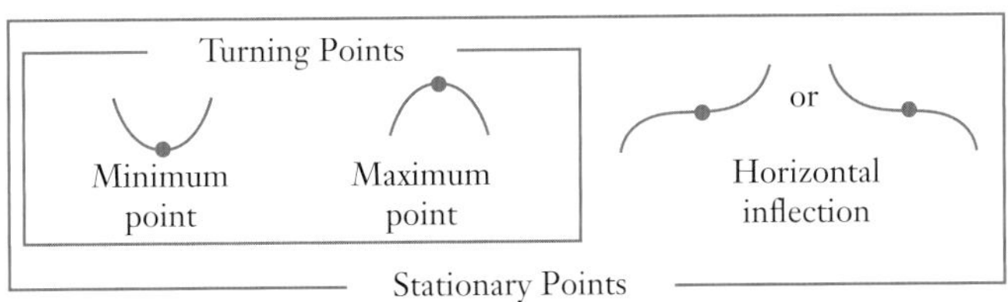

ISBN 9780170390408

Exercise 5A

1 For the function on the right, points A and I are end points, points B, D, F and H are stationary points and C, E, G and H are points of inflection.

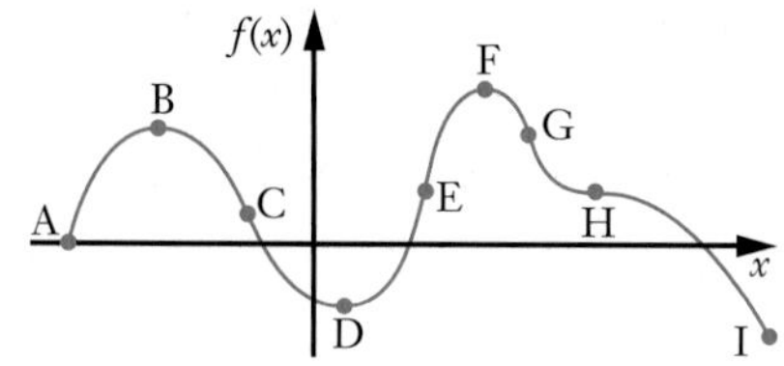

a Between which points is the function increasing, i.e. a positive gradient? (Give your answer in the form J → K, M → N, etc.)

b Between which points is the function decreasing?

c At which points is the gradient zero?

2 For each of the statements I → X state the letters of those graphs A → H for which the statement is true.

I The gradient is zero at least once.

II The gradient is always positive.

III The gradient is always negative.

IV The gradient is never negative.

V The gradient is constant.

VI The gradient is zero exactly twice.

VII For all negative x values the gradient is positive.

VIII The gradient is positive as x gets very large positively. (I.e. $x \to \infty$.)

IX The gradient is positive as x gets very large negatively. ($x \to -\infty$.)

X The gradient is negative when $x = 0$.

A

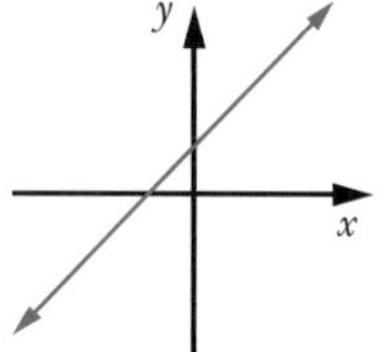

B

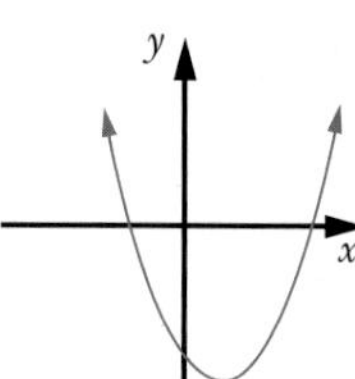

C

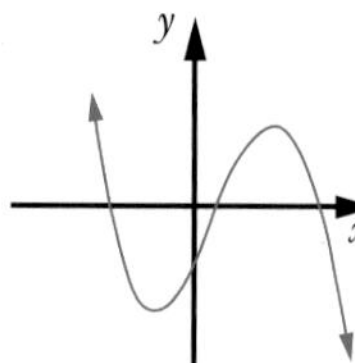

D

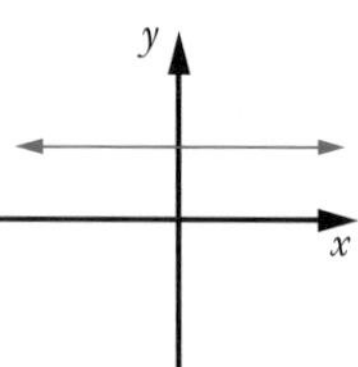

E

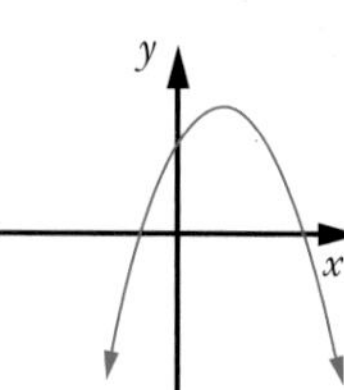

F

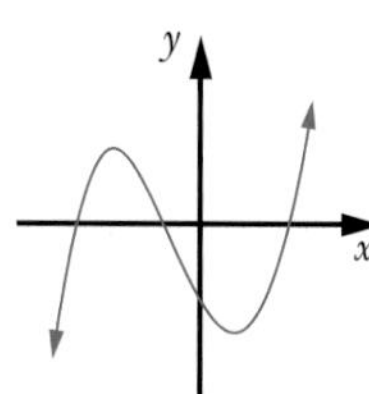

G

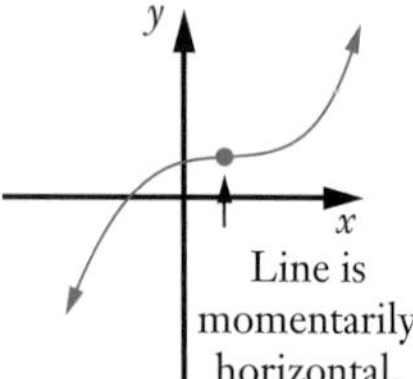

H

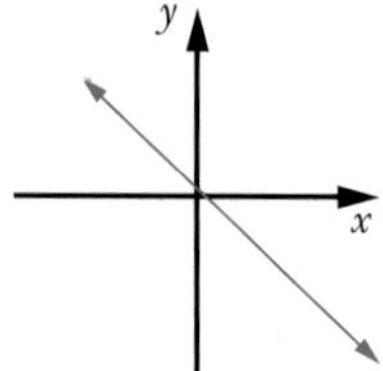

 ISBN 9780170390408

3 For the graph below state

a which *six* of the points A → P are places where the gradient is zero

b which *six* of the points A → P are places where the gradient is positive

c which *four* of the points A → P are places where the gradient is negative.

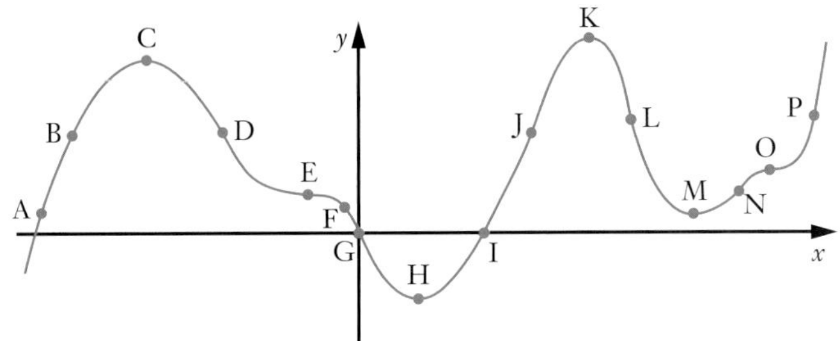

4 The diagram below shows the graph of $y = x^2$ with the tangents to the curve drawn at the point (1, 1) and the point (2, 4).

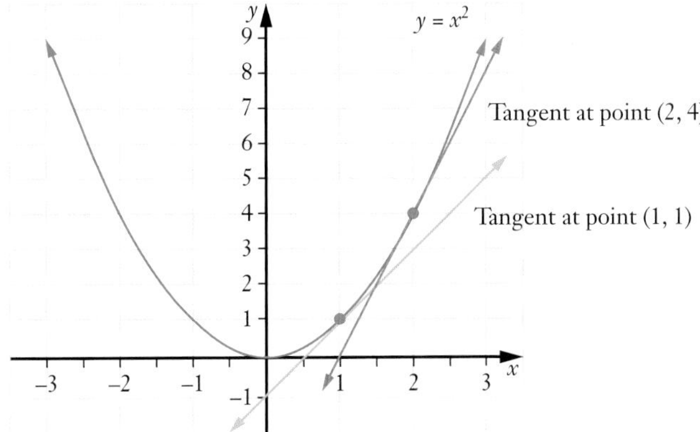

a Use the graph to suggest the gradient of $y = x^2$ at the point (1, 1).

b Use the graph to suggest the gradient of $y = x^2$ at the point (2, 4).

c Use the graph to suggest the gradient of $y = x^2$ at the point (0, 0).

d Suggest the gradient of $y = x^2$ at the point on the curve where $x = -1$.

e Suggest the gradient of $y = x^2$ at the point on the curve where $x = -2$.

f Suggest the gradient of $y = x^2 + 3$ at the point where $x = 1$.

g Suggest the gradient of $y = (x - 2)^2$ at the point where $x = 3$.

ISBN 9780170390408

5 Sketch the graph of a function that satisfies all of the conditions stated below. (You do not need to determine the equation of such a function.)

- The function cuts the x-axis at $(1, 0)$ and $(4, 0)$ and nowhere else.
- The gradient of the function is zero for $x = 2.5$.
- For $x < 2.5$ the gradient is always positive.
- For $x > 2.5$ the gradient is always negative.

6 Sketch the graph of a function that satisfies all of the conditions stated below. (You do not need to determine the equation of such a function.)

- The function cuts the x-axis at $(0, 0)$ and nowhere else.
- The gradient of the function is zero for $x = 2$.
- For $x < 2$ and for $x > 2$ the gradient is always positive.

7 Sketch the graph of a function that satisfies all of the conditions stated below. (You do not need to determine the equation of such a function.)

- The function cuts the x-axis at $(-2, 0)$, $(1, 0)$, $(6, 0)$ and nowhere else.
- The gradient of the function is zero for $x = -1$, $x = 3$ and $x = 5$.
- For $x < -1$ and for $x > 5$ the gradient is always positive.
- For $-1 < x < 3$ and $3 < x < 5$ the gradient is always negative.

Rates of change: gradients of secants

Calculating the gradient at a point on a curve

Now that we know what we mean by the gradient of a curve, how do we determine its value at various points on a curve? Well one way would be to draw the tangent to the curve at those points and estimate its gradient, as in one of the questions of the previous exercise. However, drawing the tangent accurately is difficult and deciding exactly which straight line is the tangent at a particular point involves a certain amount of guesswork. So how do we *calculate* the gradient at a particular point accurately?

To answer this question let us return to the idea mentioned after the two situations at the beginning of this chapter. It was suggested there that to determine the rate of change of the curve $y = t^3$, at the point where $t = 8$, you perhaps considered the rates of change of intervals closer and closer to $t = 8$. Let us try this approach to determine the gradient of $y = x^2$ at various points on the curve.

Consider the graph of $y = x^2$.

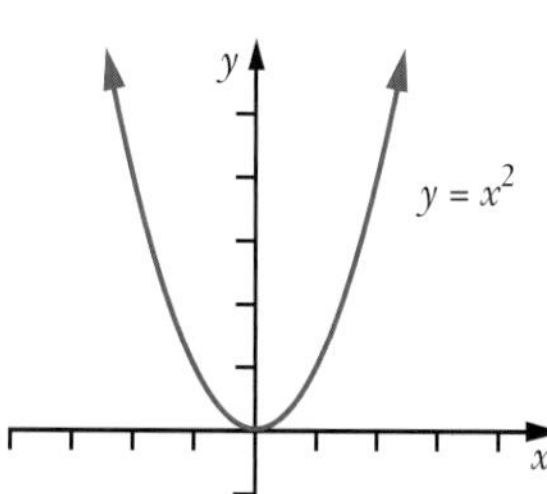

The tangent drawn through $(0, 0)$ will be the x-axis and this has a gradient of zero.

Thus the gradient of $y = x^2$ at $x = 0$ is zero but what will be the gradient of $y = x^2$ at $x = 1, 2, 3, 4, 5, \ldots$?

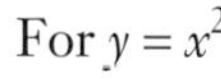
For $y = x^2$

x	0	1	2	3	4	5	...
gradient	0	?	?	?	?	?	???

ISBN 9780170390408

To determine the gradient at $x = 1$ we first determine the gradient of the **chord** PQ where P is the point (1, 1) and Q is some other point on the curve. We then move Q to positions Q_1, Q_2, Q_3, …, each position being closer to P than the previous position, and determine the gradient of the chord in each case. As Q gets closer and closer to P then so the gradient of PQ will be a better and better approximation of the gradient of the tangent at P.

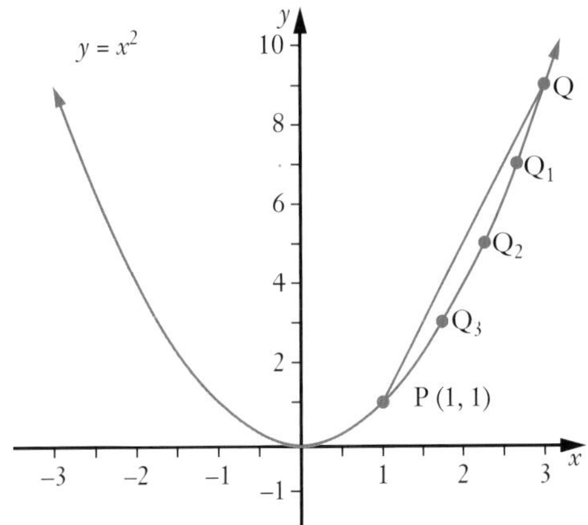

This process is shown tabulated below:

Point P	Point Q	Gradient of chord PQ
(1, 1)	(3, 9)	$\frac{9-1}{3-1} = 4$
(1, 1)	(2, 4)	$\frac{4-1}{2-1} = 3$
(1, 1)	(1.5, 2.25)	$\frac{2.25-1}{1.5-1} = 2.5$
(1, 1)	(1.1, 1.21)	$\frac{1.21-1}{1.1-1} = 2.1$
(1, 1)	(1.05, 1.1025)	$\frac{1.1025-1}{1.05-1} = 2.05$
(1, 1)	(1.01, 1.0201)	$\frac{1.0201-1}{1.01-1} = 2.01$
(1, 1)	(1.001, 1.002 001)	$\frac{1.002001-1}{1.001-1} = 2.001$

As Q approaches P the gradient of PQ approaches 2.

We say that the **limit** of the gradient of PQ, as Q approaches P, appears to be 2.

This suggests that the gradient of $y = x^2$ at $x = 1$ is 2.

Thus we now have:

For $y = x^2$

x	0	1	2	3	4	5	…
gradient	0	2	?	?	?	?	???

The first two questions of the next exercise involve determining more of the unknowns in this table.

ISBN 9780170390408

Exercise 5B

1 Complete the following table to find the gradient of $y = x^2$ at the point P(2, 4).

Point P	Point Q	Gradient of chord PQ
(2, 4)	(4, 16)	$\frac{16-4}{4-2} = ?$
(2, 4)	(3, 9)	$\frac{?-?}{?-?} = ?$
(2, 4)	(2.5, ???)	?
(2, 4)	(2.1, ???)	?
(2, 4)	(2.01, ???)	?
(2, 4)	(2.001, ???)	?
(2, 4)	(2.0001, ???)	?

Thus the gradient of $y = x^2$ at $x = 2$ is ???.

2 Repeat the 'limiting chord' process used in Question 1 to determine the gradient of $y = x^2$ at (3, 9), (4, 16) and (5, 25) and hence, together with your answer from number 1, copy and complete the following table.

For $y = x^2$

x	0	1	2	3	4	5
gradient	0	2	?	?	?	?

Use your table to suggest a rule for determining the gradient of $y = x^2$ at some point (a, a^2).

3 Repeat the 'limiting chord' process to determine the gradient of $y = 3x^2$ at (2, 12), (3, 27) and (4, 48) and hence copy and complete the table below.

For $y = 3x^2$

x	0	1	2	3	4	5
gradient	0	6	?	?	?	30

Use your table to suggest a rule for determining the gradient of $y = 3x^2$ at some point $(a, 3a^2)$.

Shutterstock.com/Guitar Photographer

ISBN 9780170390408

General statement of this 'limiting chord' process

Finding derivatives from first principles

Limits

Let us again consider the process we go through to find the gradient at a particular point, P, on a curve $y = f(x)$.

We choose some other point, Q, on the curve whose x-coordinate is a little more than that of point P. Suppose P has an x-coordinate of x and Q has an x-coordinate of $(x + h)$.

The corresponding y-coordinates of P and Q will then be $f(x)$ and $f(x + h)$.

Thus the gradient of PQ $= \dfrac{f(x+h)-f(x)}{h}$.

This gives us the **average rate of change** of the function from P to Q.

For example the **average rate of change** of the function $y = x^2$ from P(3, 9) to Q(4, 16) is given by $\dfrac{16-9}{4-3} = 7$.

We then bring Q closer and closer to P, i.e. we allow h to tend to zero, and we determine the limiting value of the gradient of PQ.

i.e. Gradient at P = limit of $\dfrac{f(x+h)-f(x)}{h}$ as h tends to zero.

We write this as:

$$\text{Gradient at } P(x, f(x)) = \lim_{h\to 0}\frac{f(x+h)-f(x)}{h}$$

This gives us the **instantaneous rate of change** of the function at P using an algebraic approach, rather than having to create tables as we did earlier.

For example the **instantaneous rate of change** of the function $y = x^2$ at the point P(3, 9) is given by

$$\begin{aligned}\lim_{h\to 0}\frac{f(3+h)-f(3)}{h} &= \lim_{h\to 0}\frac{(3+h)^2-(3)^2}{h}\\ &= \lim_{h\to 0}\frac{9+6h+h^2-9}{h}\\ &= \lim_{h\to 0}\frac{6h+h^2}{h}\\ &= \lim_{h\to 0}(6+h)\\ &= 6 \qquad \text{because } h\to 0 \text{ then so } (6+h)\to 6\end{aligned}$$

Does this agree with the answer you obtained numerically in Question 2 of the previous exercise?

This algebraic method for determining the gradient at a point on $f(x) = x^2$ is certainly a quicker process than creating the tables that we did earlier. However, rather than using this quicker algebraic method each time we want to determine the instantaneous rate of change at some particular point on $y = x^2$, we could instead apply the technique *once* for the general point (x, x^2), obtain a formula for the gradient, and then apply this formula each time.

Consider the general point $P(x, x^2)$ lying on the function $f(x) = x^2$.

Applying the general result obtained previously:

$$\begin{aligned}\text{Gradient at } P(x, x^2) &= \lim_{h \to 0} \frac{(x+h)^2 - x^2}{h} \\ &= \lim_{h \to 0} \frac{x^2 + 2xh + h^2 - x^2}{h} \\ &= \lim_{h \to 0} \frac{2xh + h^2}{h} \\ &= \lim_{h \to 0} (2x + h) \\ &= 2x \qquad \text{because as } h \to 0 \text{ then so } (2x+h) \to 2x.\end{aligned}$$

Thus for the curve $y = x^2$ the gradient formula or gradient function is $2x$.

Does this agree with your answers and suggested rule for Exercise 5B Question 2?

This process of determining the gradient formula or gradient function of a curve or function is called **DIFFERENTIATION** (part of the branch of mathematics known as **CALCULUS**).

If we **differentiate** $y = x^2$ with respect to the variable x, we obtain the **gradient function** $2x$.

We say that $2x$ is the **derivative** of x^2.

Similarly:

- If we differentiate $y = t^2$ with respect to the variable t we obtain the gradient function $2t$.
- If we differentiate $z = y^2$ with respect to the variable y we obtain the gradient function $2y$.
- If we differentiate $v = z^2$ with respect to the variable z we obtain the gradient function $2z$, etc.

Shutterstock.com/Curloso

 ISBN 9780170390408

Exercise 5C

On the previous page the result

$$\text{Gradient at } P(x, f(x)) = \lim_{h \to 0} \frac{f(x+h) - f(x)}{h}$$

was used to determine the gradient function of $y = x^2$ as $2x$.

Use this same procedure to prove the following results.

1 The gradient function of $y = 4x^2$ is $8x$.

2 The gradient function of $y = 2x^3$ is $6x^2$.

3 The gradient function of $y = x^4$ is $4x^3$.

The results given in **Exercise 5C**, and that you should have found in **Exercise 5B**, suggest that if $y = ax^n$ then the gradient function is anx^{n-1}.

In words this general statement can be remembered as:

'multiply by the power and decrease the power by one'

This 'suggested' general statement is indeed true but can we **prove** it? Well, to do so we simply have to go back to the basic principle that

$$\text{Gradient at P}(x, f(x)) = \lim_{h \to 0} \frac{f(x+h) - f(x)}{h}$$

and apply it to the function $y = ax^n$.

However, before turning the page and seeing it done for you, try it yourself first.

One result that you may find useful is the *binomial expansion*, a result you were reminded of in the *Preliminary work* section at the beginning of this book:

$$(p + q)^n = p^n + {}^nC_1 p^{n-1}q^1 + {}^nC_2 p^{n-2}q^2 + {}^nC_3 p^{n-3}q^3 + \ldots + {}^nC_n p^0 q^n$$

An alternative approach would be to use another result that was mentioned in the *Preliminary work*:

$$p^n - q^n = (p - q)(p^{n-1} + p^{n-2}q + p^{n-3}q^2 + p^{n-4}q^3 + \ldots pq^{n-2} + q^{n-1})$$

Have a go!

ISBN 9780170390408

Consider some general point $P(x, ax^n)$ on $f(x) = ax^n$.

$$\begin{aligned}
\text{The gradient at } P(x, ax^n) &= \lim_{h \to 0} \frac{f(x+h) - f(x)}{h} \\
&= \lim_{h \to 0} \frac{a(x+h)^n - ax^n}{h} \qquad [1] \\
&= \lim_{h \to 0} \frac{a\left((x+h)^n - x^n\right)}{h} \\
&= \lim_{h \to 0} \frac{a\left(x^n + {}^nC_1x^{n-1}h + {}^nC_2x^{n-2}h^2 + \ldots + h^n - x^n\right)}{h} \\
&= \lim_{h \to 0} \frac{a\left({}^nC_1x^{n-1}h + {}^nC_2x^{n-2}h^2 + \ldots + h^n\right)}{h} \\
&= \lim_{h \to 0} \left(a\,{}^nC_1x^{n-1} + a\,{}^nC_2x^{n-2}h + a\,{}^nC_3x^{n-3}h^2 \ldots + ah^{n-1}\right) \\
&= a\,{}^nC_1x^{n-1} \\
&= anx^{n-1}
\end{aligned}$$

The reader is left to show that the same result can be arrived at by applying the rule given at the bottom of the previous page to equation [1] above.

Notation

In the expression

$$\lim_{h \to 0} \frac{f(x+h) - f(x)}{h},$$

'h' is a small **increment** in the variable x and

$$[f(x+h) - f(x)]$$

is the corresponding small **increment** in the variable y.

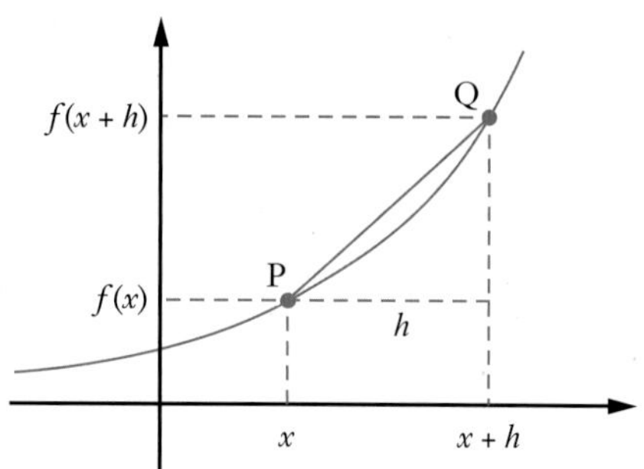

Denoting this small increment in x as δx, where 'δ' is a Greek letter pronounced 'delta', and the small increment in y as δy, we have:

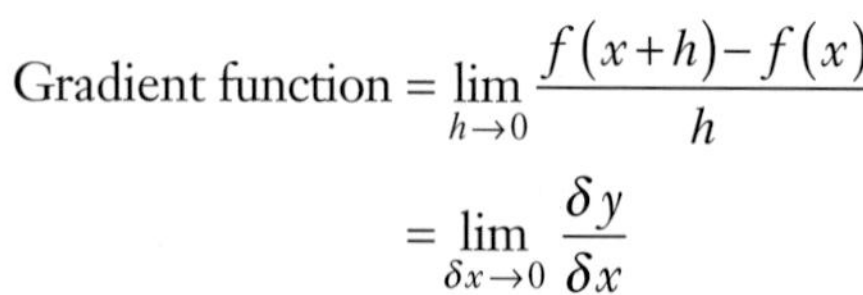

$$\begin{aligned}
\text{Gradient function} &= \lim_{h \to 0} \frac{f(x+h) - f(x)}{h} \\
&= \lim_{\delta x \to 0} \frac{\delta y}{\delta x}
\end{aligned}$$

which we write as $\frac{dy}{dx}$ (pronounced 'dee y by dee x').

ISBN 9780170390408

Thus if $y = x^2$, then $\frac{dy}{dx}$, the gradient function, is $2x$.

If $y = x^3$, then $\frac{dy}{dx}$, the gradient function, is $3x^2$.

If $y = x^4$, then $\frac{dy}{dx}$, the gradient function, is $4x^3$.

> If $y = ax^n$ then $\frac{dy}{dx}$, the gradient function, is anx^{n-1}.

This can also be written as $\frac{d}{dx}(y) = \frac{d}{dx}(ax^n)$

$= anx^{n-1}$.

We say 'dee by dee x' of ax^n is anx^{n-1}. I.e. the derivative of ax^n is anx^{n-1}.

EXAMPLE 1

Determine the gradient function for each of the following.

a $y = 3x^2$

b $y = 7x^3$

c $y = 2x^5$

d $y = 3x$

e $y = 7$

Solution

a If $y = 3x^2$ then

$$\frac{dy}{dx} = 3(2)x^{2-1} = 6x$$

b If $y = 7x^3$ then

$$\frac{dy}{dx} = 7(3)x^{3-1} = 21x^2$$

c If $y = 2x^5$ then

$$\frac{dy}{dx} = 2(5)x^{5-1} = 10x^4$$

d If $y = 3x$ (i.e. $3x^1$) then

$$\frac{dy}{dx} = 3(1)x^{1-1} = 3$$

(as expected because $y = 3x$ is a straight line with gradient 3)

e If $y = 7$ (i.e. $7x^0$) then

$$\frac{dy}{dx} = 7(0)x^{0-1} = 0$$

(as we would expect because $y = 7$ is a horizontal line)

The derivatives can also be obtained from some calculators.

$\frac{d}{dx}(3x^2)$	$6\cdot x$
$\frac{d}{dx}(7x^3)$	$21\cdot x^2$
$\frac{d}{dx}(2x^5)$	$10\cdot x^4$
$\frac{d}{dx}(3x)$	3
$\frac{d}{dx}(7)$	0

ISBN 9780170390408

EXAMPLE 2

Determine the gradient of the curve $y = 3x^4$ at the point (2, 48).

Solution

If $\quad y = 3x^4$

then $\quad \dfrac{dy}{dx} = 12x^3.$

At (2, 48), $x = 2$.

Thus $\quad \dfrac{dy}{dx} = 12\,(2)^3$

$\quad = 96.$

The gradient of $y = 3x^4$ at (2, 48) is 96.

$\frac{d}{dx}(3\cdot x^4 \mid x=2)$ 96

TECHNOLOGY

Get to know the capability of your calculator with regard to finding the derivative of a function and of finding the value of the derivative for a specific x value. However make sure that if the course requires it you can also determine derivatives, and gradients at a point, yourself, without access to a calculator.

Note: For the moment we are differentiating functions of the form $y = ax^n$ for n a non-negative integer. Later in this chapter we will consider more general **polynomial functions** which, as the reader should know, are of the form

$$f(x) = a_n x^n + a_{n-1}x^{n-1} + a_{n-2}x^{n-2} + \ldots + a_2x^2 + a_1x + a_0$$

where n is a non-negative integer and $a_n, a_{n-1}, a_{n-2}, \ldots$ are all numbers, called the **coefficients** of x^n, x^{n-1}, x^{n-2} etc.

The highest power of x is the **order** of the polynomial.

Thus **linear functions,** $y = mx + c$, are polynomials of order 1,
quadratic functions, $y = ax^2 + bx + c$, are polynomials of order 2,
cubic functions, $y = ax^3 + bx^2 + cx + d$, are polynomials of order 3, and so on.

Shutterstock.com/Matt Gore

ISBN 9780170390408

EXAMPLE 3

Find the coordinates of any points on the curve $y = x^3$ where the gradient is 12.

Solution

If $y = x^3$ then $\frac{dy}{dx} = 3x^2$

Thus we require points for which $3x^2 = 12$,

i.e. $x^2 = 4$

giving $x = 2$ or $x = -2$

If $x = 2, y = 2^3 = 8$ and if $x = -2, y = (-2)^3 = -8$

Thus $y = x^3$ has a gradient of 12 at (2, 8) and (−2, −8).

Note:

- If $y = f(x)$ then the derivative of y with respect to x can be written as $\frac{dy}{dx}, \frac{df}{dx}$ or $\frac{d}{dx}f(x)$.
  (This last version is pronounced: '*Dee by dee x of eff of x*'.)
- A shorthand notation using a 'dash' is sometimes used for differentiation with respect to x.
  Thus if $y = f(x)$ we can write $\frac{dy}{dx}$ as $f'(x)$ or simply y' or f'.

EXAMPLE 4

Determine $f'(x)$ for

a $f(x) = 7x^5$ b $f(x) = 20$ c $f(x) = 6x^9$

Solution

a If $f(x) = 7x^5$, then $f'(x) = 35x^4$

b If $f(x) = 20$, then $f'(x) = 0$

c If $f(x) = 6x^9$, then $f'(x) = 54x^8$

Shutterstock.com/Gray wall studio

Slopes of curves

Finding the equation of a tangent at a point on $y = ax^n$

EXAMPLE 5

Find the equation of the tangent to $y = 0.5x^3$ at the point (2, 4).

Solution

Find the derivative either algebraically or by calculator.

If $y = 0.5x^3$ then $\frac{dy}{dx} = 1.5x^2$

Thus at (2, 4), $\frac{dy}{dx} = 1.5(2)^2$

$= 6$

$\frac{d}{dx}(0.5x^3)|x=2$ 6

Thus the gradient of the curve $y = 0.5x^3$, at the point (2, 4), is 6.

Thus the gradient of the tangent to $y = 0.5x^3$, at the point (2, 4), is also 6.

The tangent is a straight line and has an equation of the form $y = 6x + c$.
But (2, 4) lies on this tangent $\therefore\ 4 = 6(2) + c$
giving $c = -8$

The tangent to $y = 0.5x^3$ at the point (2, 4) has equation $y = 6x - 8$.

Some calculators and internet programs are able to determine the equation of a tangent at a point on a curve directly, given the appropriate instructions. Whilst you are encouraged to explore this capability of such programs make sure you can carry out the process shown in the above example yourself.

Though we have been concentrating on finding the gradients at points on various curves it is important to remember that the gradient tells us the rate at which one variable is changing with respect to another. Rates of change are important in everyday life.

Differentiation can be used to find:

- the rate at which a vehicle is changing its position with respect to time, i.e. the vehicle's speed.
- the rate of change in the population of a country.
- the rate of change in the number of people suffering a disease.
- the rate of change in the value of one currency with respect to another.
- the rate of change in the total profit we get from a particular item with respect to the unit cost of that item.

Etc.

ISBN 9780170390408

In the remainder of this chapter we will concentrate on improving our ability to differentiate various functions. In the next chapter we will apply these skills to some real-life rate of change situations.

Exercise 5D

Determine the gradient function $\frac{dy}{dx}$ for each of the following.

1 $y = x^2$

2 $y = x^3$

3 $y = x$

4 $y = x^4$

5 $y = 3$

6 $y = 6x^2$

7 $y = 6x^4$

8 $y = 7x$

9 $y = 16x$

10 $y = 2x^7$

11 $y = 7x^2$

12 $y = 9x$

13 $y = \frac{x^2}{10}$

14 $y = \frac{2x^6}{3}$

15 $y = \frac{3x^6}{2}$

16 $y = \frac{2x^7}{7}$

Differentiate each of the following with respect to x.

17 $4x^2$

18 $5x^4$

19 $8x^3$

20 9

21 x^7

22 $4x^6$

23 $9x^2$

24 $5x$

Determine $f'(x)$ for each of the following.

25 $f(x) = 5$

26 $f(x) = 6x^3$

27 $f(x) = 8x^4$

28 $f(x) = 3x^5$

29 $f(x) = x^6$

30 $f(x) = 6x^7$

31 $f(x) = 4x^4$

32 $f(x) = 10x$

Determine the gradient of each of the following at the given point.

33 $y = 2x^2$ at the point (3, 18)

34 $y = 4x^3$ at the point (1, 4)

35 $y = 4x^3$ at the point (−1, −4)

36 $y = x^5$ at the point (2, 32)

37 $y = 7x$ at the point (2, 14)

38 $y = 5x^2$ at the point (−2, 20)

39 $y = 0.25x^2$ at the point (4, 4)

40 $y = \frac{x^2}{5}$ at the point (2, 0.8)

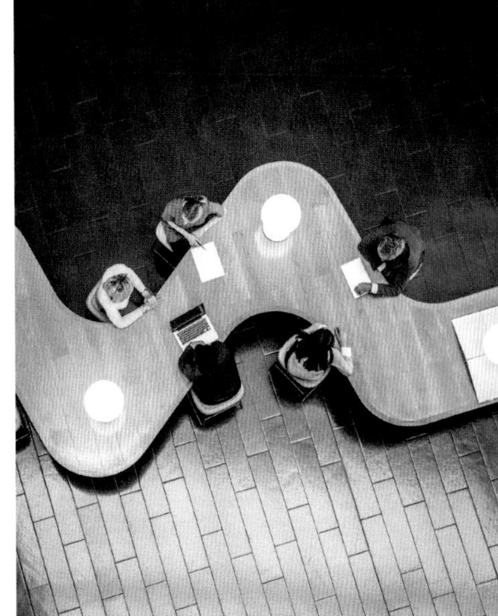

iStock.com/JohnnyGreig

ISBN 9780170390408

Find the coordinates of the point(s) on the following curves where the gradient is as stated.

41 $y = x^4$, gradient 4.

42 $y = x^3$, gradient 3.

43 $y = 3x^2$, gradient 9.

44 $y = 2x^3$, gradient 1.5.

45 $y = x^6$, gradient 6.

46 $y = x^6$, gradient –6.

Find the equation of the tangent to the following curves at the indicated point.

47 $y = 2x^3$ at the point (1, 2)

48 $y = 3x^2$ at the point (–1, 3)

49 $y = 5x^2$ at the point (2, 20)

50 $y = 5x^2$ at the point (–2, 20)

51 $y = \frac{x^4}{2}$ at the point (2, 8)

52 $y = \frac{x^3}{6}$ at the point (6, 36)

53 If $f(x) = 3x^3$, find

a $f(2)$ **b** $f(-1)$ **c** $f'(x)$ **d** $f'(2)$

54 If $f(x) = 1.5x^2$, find

a $f(2)$ **b** $f(4)$ **c** $f'(x)$ **d** $f'(2)$

55 For $y = 2x^3$, determine:

a by how much y changes when x changes from $x = 2$ to $x = 5$.

b the average rate of change in y, per unit change in x, when x changes from $x = 2$ to $x = 5$.

c the instantaneous rate of change of y, with respect to x, when $x = 2$.

d the instantaneous rate of change of y, with respect to x, when $x = 5$.

56 The straight line $y = 8x + 16$ cuts the curve $y = 8x^2$ at two points. Find the coordinates of each point and the gradient of the curve at each one.

57 The straight line $y = 4x$ cuts the curve $y = x^3$ at three points. Find the coordinates of each point and the gradient of the curve at each one.

58 The tangent to the curve $y = ax^4$ at the point $(3, b)$ has a gradient of 2. Find the values of a and b.

59 The tangent to the curve $y = ax^3$ at the point $(-1, b)$ is perpendicular to the line $y = 2x + 3$. Find the values of a and b.

(As mentioned in the *Preliminary work*: If two lines are perpendicular the product of their gradients is –1.)

ISBN 9780170390408

Differentiating $f(x) \pm g(x)$

Derivatives of polynomials

Basic differentiation

Derivative of a sum of terms

Derivatives of linear products

Tangents to a curve

We know that if $y = 3x$ then $\dfrac{dy}{dx} = 3$

and if $y = x^2$ then $\dfrac{dy}{dx} = 2x$.

It then seems reasonable to suggest that if $y = 3x + x^2$

then $\dfrac{dy}{dx} = 3 + 2x$.

To check whether this seemingly reasonable suggestion is true, we differentiate $3x + x^2$ from first principles, i.e. by determining $\lim\limits_{h \to 0} \dfrac{f(x+h) - f(x)}{h}$.

Let $\quad f(x) = 3x + x^2$,

then $\quad f(x+h) = 3(x+h) + (x+h)^2$.

$$\text{Thus } \lim_{h \to 0} \frac{f(x+h)-f(x)}{h} = \lim_{h \to 0} \frac{[3(x+h)+(x+h)^2]-(3x+x^2)}{h}$$

$$= \lim_{h \to 0} \frac{3x+3h+x^2+2xh+h^2-3x-x^2}{h}$$

$$= \lim_{h \to 0} (3+2x+h)$$

$$= 3+2x$$

Though the above example only considers the particular function $(3x + x^2)$, it is in fact true that

If $\quad y = f(x) \pm g(x)$

then $\quad \dfrac{dy}{dx} = f'(x) \pm g'(x)$

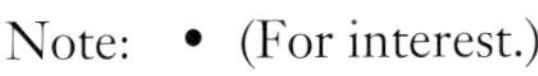

Note:
- (For interest.)
  If some operation, which we will call Z, applied to the functions f and g is such that
  $$Z(f+g) = Z(f) + Z(g)$$
  and $\quad Z(k \times f) = k \times Z(f) \quad$ for k a constant
  then Z is said to be a *linear* operator.

  Hence the facts that $\quad \dfrac{d}{dx}(ky) = k\dfrac{dy}{dx} \quad$ for k a constant

  and $\quad \dfrac{d}{dx}(y_1 + y_2) = \dfrac{dy_1}{dx} + \dfrac{dy_2}{dx}$

  confirm what may be referred to as the 'linearity property' of derivatives.
- (For interest.)
  The use of δx and δy to represent small increments in x and y respectively, and the use of the term $\dfrac{dy}{dx}$, is referred to as 'Leibniz notation' in honour of the German mathematician Gottfried Leibniz (1646–1716).

ISBN 9780170390408

EXAMPLE 6

Find the gradient of $y = x^2 - 3x$ at the point (5, 10).

Solution

By calculator (typical display below):

```
d
── (x²–3·x) | x=5
dx
                         7
```

Algebraically:

If $y = x^2 - 3x$,

then $\dfrac{dy}{dx} = 2x - 3$.

Therefore, at the point (5, 10),

$$\frac{dy}{dx} = 2(5) - 3$$
$$= 7.$$

The gradient of $y = x^2 - 3x$ at (5, 10) is 7.

Exercise 5E

Find the gradient function $\dfrac{dy}{dx}$ for each of the following.

1 $y = x^2 + 3x$

2 $y = x^3 - 4x + 7$

3 $y = 6x^2 - 7x^3 + 4$

4 $y = 3x^4 + 2x^3 - 5x$

5 $y = 6 + 7x + x^2$

6 $y = 6x^2 - 3x$

7 $y = 4x^2 + 7x - 1$

8 $y = 5x^3 - 4x^2 + 8$

9 $y = 5x^4 - 3x + 11$

10 $y = 2x^2 + 7x + 1$

11 $y = 5 - 3x^2 + 7x$

12 $y = 1 + x + x^2 + x^3 + x^4$

13 $y = 5 - 4x + 3x^2 - 2x^3 + x^4$

14 Find the gradient of $y = x^3 - 3x^2$ at the point (1, –2).

15 Find the gradient of $y = 17 + 2x^3$ at the point (–2, 1).

16 Find the gradient of $y = x^3 - x^2 - 8$ at the point (3, 10).

17 Find the gradient of $y = 1 + 3x - 2x^3 + x^4$ at the point (2, 7).

18 Find the equation of the tangent to $y = x^2 + 3x$ at the point (2, 10).

19 Find the equation of the tangent to $y = 2x^2 - 7x$ at the point (5, 15).

20 Find the equation of the tangent to $y = x^3 - 5x^2 + 14$ at the point (4, –2).

21 Find the equation of the tangent to $y = 5x^4 - 4x^5$ at the point (1, 1).

22 Find the coordinates of any point on the curve $y = x^3 + 6x^2 - 10x + 1$ where the gradient is 5.

23 The curve $y = x^2 - 2x - 15$ cuts the x-axis in two places. Find the coordinates of each of these points and determine the gradient of the curve at each one.

ISBN 9780170390408

24 Find the coordinates of any point on the curve $y = x^2 - 7x$ where the gradient is the same as that of $3y = 9x - 1$.

25 Find the coordinates of any point on the curve $y = x^3 + 3x^2 - 7x - 1$ where the gradient is the same as that of $y = 2x + 3$.

Differentiating more general power functions

Functions of the form $y = ax^n$ are called **power functions** and in this chapter we have considered such functions for non-negative integer values of n. We have differentiated such functions using the fact that

$$\text{if } y = ax^n \text{ then } \frac{dy}{dx} = anx^{n-1}.$$

Polynomial functions, which are linear combinations of such power functions, could then be differentiated using the fact that

$$\text{if } y = f(x) \pm g(x) \text{ then } \frac{dy}{dx} = f'(x) \pm g'(x).$$

Now let us remove the restriction that n must be a non-negative integer and consider more general power functions $y = ax^n$, for example $y = \sqrt{x}$ and $y = \frac{1}{x}$.

Though not proved here, it is the case that for negative and fractional values of n, the same rule applies, i.e,

$$\text{If } y = ax^n \text{ then } \frac{dy}{dx} = anx^{n-1}.$$

Thus if $y = \frac{1}{x}$ i.e. $y = x^{-1}$,

$$\text{then } \frac{dy}{dx} = -1x^{-2}$$
$$= -\frac{1}{x^2},$$

and if $y = \sqrt{x}$ i.e. $y = x^{\frac{1}{2}}$,

$$\text{then } \frac{dy}{dx} = \frac{1}{2}x^{-\frac{1}{2}}$$
$$= \frac{1}{2\sqrt{x}}.$$

$$\frac{d}{dx}\left(\frac{1}{x}\right) \qquad \frac{-1}{x^2}$$
$$\frac{d}{dx}\left(\sqrt{x}\right) \qquad \frac{1}{2\cdot\sqrt{x}}$$

It then follows that:

If $$y = 3x^2 - 2x + 7 + \sqrt[3]{x^2} - \frac{5}{x^2}$$

i.e. $$y = 3x^2 - 2x + 7 + x^{\frac{2}{3}} - 5x^{-2}$$

$$\frac{dy}{dx} = 6x - 2 + \frac{2}{3}x^{-\frac{1}{3}} + 10x^{-3}$$
$$= 6x - 2 + \frac{2}{3\sqrt[3]{x}} + \frac{10}{x^3}$$

ISBN 9780170390408

EXAMPLE 7

Find the gradient of $y = x^2 + \frac{16}{x}$ at the point (4, 20).

Solution

By calculator (typical display below).

```
d  (    16)
-- (x² + --) | x=4
dx (    x )
                        7
```

Algebraically:

$$\text{If } y = x^2 + \frac{16}{x}$$

$$= x^2 + 16x^{-1}$$

$$\text{then } \frac{dy}{dx} = 2x - 16x^{-2}$$

Therefore, at the point (4, 20),

$$\frac{dy}{dx} = 2(4) - 16(4)^{-2}$$

$$= 7$$

The gradient of $y = x^2 + \frac{16}{x}$ at (4, 20) is 7.

EXAMPLE 8

Find the equation of the tangent to $y = 12\sqrt{x}$ at the point (4, 24).

Solution

Either algebraically: or by calculator:

If $y = 12x^{\frac{1}{2}}$ then $\frac{dy}{dx} = 6x^{-\frac{1}{2}}$

Thus at (4, 24), $\frac{dy}{dx} = 6(4)^{-\frac{1}{2}}$

$$= \frac{6}{\sqrt{4}}$$

$$= 3$$

```
d
--(12√x) | x=4
dx
                        3
```

We determine that the gradient of the curve $y = 12\sqrt{x}$, at the point (4, 24), is 3.

Thus the gradient of the tangent to $y = 12\sqrt{x}$, at the point (4, 24), is also 3.

The tangent, being a straight line, will have equation of the form $y = 3x + c$.

But (4, 24) lies on this tangent $\therefore\ 24 = 3(4) + c$

giving $c = 12$

The required tangent has equation $y = 3x + 12$.

ISBN 9780170390408

Exercise 5F

(Whilst you are encouraged to use your calculator to obtain expressions for the derivative, and to determine its value at particular points on a curve, it is suggested that you do most of the following questions algebraically to ensure that you can follow the basic processes without a calculator.)

Determine the gradient function $\frac{dy}{dx}$ for each of the following.

1 $y = \sqrt{x}$

2 $y = \frac{1}{x}$

3 $y = \frac{3}{x}$

4 $y = 6x^{\frac{1}{2}}$

5 $y = 6x^{\frac{1}{3}}$

6 $y = \sqrt{x^3}$

7 $y = 2\sqrt[3]{x}$

8 $y = \frac{1}{x^3}$

9 $y = \frac{1}{x^4}$

10 $y = \frac{2}{x^3}$

11 $y = \frac{5}{x^4}$

12 $y = x^2 + \sqrt{x}$

13 $y = 3x^2 - 4\sqrt{x}$

14 $y = x + \frac{1}{x}$

15 $y = x^2 - \frac{1}{x^2}$

16 $y = \sqrt{x} + \frac{3}{x}$

17 $y = x^2 + x + 1 + \frac{1}{x} + \frac{1}{x^2}$

Determine $f'(x)$ for each of the following.

18 $f(x) = \frac{2}{x}$

19 $f(x) = \frac{3}{\sqrt{x}}$

20 $f(x) = \frac{6}{\sqrt[3]{x}}$

21 $f(x) = \frac{1}{\sqrt[3]{x}}$

22 Find the gradient of $y = \frac{4}{x} - x^2$ at the point $(2, -2)$.

23 Find the gradient of $y = \frac{1}{x^2}$ at the point $\left(-2, \frac{1}{4}\right)$.

24 Find the gradient of $y = 1 - \frac{1}{x}$ at the point $(4, 0.75)$.

25 Find the gradient of $y = 3x^3 - \frac{2}{x}$ at the point $(1, 1)$.

26 Find the gradient of $y = \sqrt[3]{(x^4)}$ at the point (8, 16).

27 Find the gradient of $y = 6\sqrt[3]{x} + \frac{2}{x^3}$ at the point (1, 8).

28 Find the gradient of $y = \frac{2}{x} + x^2 + \frac{16}{x^2}$ at the point (2, 9).

29 Find the coordinates of the point(s) on the curve $y = \frac{1}{x}$ where the gradient is equal to $-\frac{1}{4}$.

30 Find the coordinates of the point(s) on the curve $y = \sqrt{x}$ where the gradient is equal to 1.

31 Find the coordinates of any point on the curve $y = x^2 - 108\sqrt{x}$ where the gradient is zero.

32 Find the equation of the tangent to the curve $y = \sqrt{x}$ at the point (4, 2).

33 Find the equation of the tangent to the curve $y = \frac{1}{x}$ at the point (1, 1).

34 Find the equation of the tangent to the curve $y = \frac{1}{x^2}$ at the point (2, 0.25).

35 Find the coordinates of any point on the curve $y = 2x - \frac{1}{x}$ where the gradient is the same as that of $16y = 41x + 6$.

36 (Challenge)
Use the first principles definition

$$\text{Gradient at } P(x, f(x)) = \lim_{h \to 0} \frac{f(x+h) - f(x)}{h}$$

to show that if $y = \frac{1}{x}$, then $\frac{dy}{dx} = -\frac{1}{x^2}$

and if $y = \sqrt{x}$, then $\frac{dy}{dx} = \frac{1}{2\sqrt{x}}$.

ISBN 9780170390408

Miscellaneous exercise five

This miscellaneous exercise may include questions involving the work of this chapter, the work of any previous chapters, and the ideas mentioned in the Preliminary work section at the beginning of the book.

1 Find the value of n in each of the following.

a $5 \times 5 \times 5 \times 5 = 5^n$ **b** $2^4 = n$ **c** $2^n = 8$

d $6^3 \times 6^4 = 6^n$ **e** $2^6 \times 8 = 2^n$ **f** $3^2 \times 3^n = 3^6$

g $100 \times 10^n = 10^6$ **h** $16 \times 8 = 2^n$ **i** $4 \times 16 = 4^n$

j $8^5 \div 8^n = 8^2$ **k** $15^n = 1$ **l** $3^2 \times 3^n \times 3 = 3^7$

m $5^9 \div 5^3 \times 5^n = 5^8$ **n** $5^9 \div (5^3 \times 5^n) = 5^2$ **o** $8 \times 8 \times 8 = 2^n$

2 a Find the average rate of change of the function $y = x^2$ from the point P(4, 16) to the point Q(5, 25).

b Find the instantaneous rate of change of the function $y = x^2$ at the point with coordinates (8, 64).

3 a Find the average rate of change of the function $y = x^3$ from the point on the curve where $x = 1$ to the point on the curve where $x = 3$.

b Find the instantaneous rate of change of the function $y = -2x^3$ at the point on the curve with coordinates (–2, 16).

4 If we consider our 'ancestors' to be our parents, grandparents, great grandparents, great great grandparents etc. then going back one generation we each have two ancestors in that generation, going back two generations we each have four ancestors in that generation, etc. (assuming no repetition of ancestors). How many ancestors does a person have

Shutterstock.com/Monkey Business Images

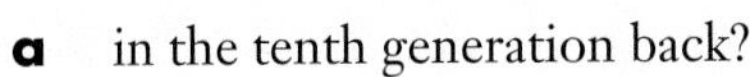

a in the tenth generation back?

b in the thirtieth generation back?

How many ancestors does a person have altogether if we sum the ancestors from

c the first generation back to the tenth generation back?

d the first generation back to the thirtieth generation back?

5 Differentiate each of the following with respect to x.

a $5 - x^3$ **b** $5x^2 - 6\sqrt{x}$ **c** $5x^2 + 6 + \dfrac{1}{x^2}$

6 Find the gradient of $y = \dfrac{1}{x}$ at the point (0.5, 2).

ISBN 9780170390408

7 For each of the tables shown below determine whether the relationship that exists between x and y is:

- linear, (rule can be written in the form $y = mx + c$),
- quadratic, (rule can be written in the form $y = ax^2 + bx + c$),
- cubic, (rule can be written in the form $y = ax^3 + bx^2 + cx + d$),
- exponential, (rule can be written in the form $y = a \times b^x$),
- reciprocal, (rule can be written in the form $y = \frac{k}{x}$),

or none of the above five types.

For those that are one of the above five types determine the algebraic rule for the relationship in the form 'y = ???'.

('undef' indicates that the function is undefined for that value of x.)

a

x	−4	−3	−2	−1	0	1	2	3	4
y	1.5	2	3	6	undef	−6	−3	−2	−1.5

b

x	−4	−3	−2	−1	0	1	2	3	4
y	17	10	5	2	1	2	5	10	17

c

x	−4	−3	−2	−1	0	1	2	3	4
y	−7	−4	−1	2	5	8	11	14	17

d

x	−4	−3	−2	−1	0	1	2	3	4
y	0.0016	0.008	0.04	0.2	1	5	25	125	625

e

x	−4	−3	−2	−1	0	1	2	3	4
y	12	6	2	0	0	2	6	12	20

Hint: To obtain the rule for part **e** consider the y values as:

$-4 \times -3 \quad -3 \times -2 \quad -2 \times -1 \quad -1 \times 0 \quad 0 \times 1 \quad 1 \times 2 \quad 2 \times 3 \quad 3 \times 4 \quad 4 \times 5$

f

x	−4	−3	−2	−1	0	1	2	3	4
y	0.0001	0.001	0.01	0.1	1	10	100	1000	10 000

g

x	−4	−3	−2	−1	0	1	2	3	4
y	0.25	0.5	1	2	4	8	16	32	64

h

x	−4	−3	−2	−1	0	1	2	3	4
y	6	8	12	24	undef	−24	−12	8	−6

i

x	−4	−3	−2	−1	0	1	2	3	4
y	−56	0	20	16	0	−16	−20	0	56

ISBN 9780170390408

8 A triangle has its three angles in arithmetic progression. If the smallest angle is 10° find the size of the other two angles.

9 A particular sequence is geometric with a common ratio of 5 and a fourth term equal to 100. Define the sequence by stating the first term, T_1, and giving T_{n+1} in terms of T_n.

10 Given that $a = 2 \times 10^7$ and $b = 4 \times 10^4$ evaluate each of the following, without the assistance of a calculator, giving your answers in standard form (scientific notation).

a $a \times b$ **b** $b \times a$ **c** a^3 **d** b^2 **e** $b \div a$ **f** $a \div b$

11 For each of the following sequences:

Sequence 1: $T_1 = 5$ and $T_{n+1} = 3T_n + 2$

Sequence 2: $T_1 = 0.125$ and $T_{n+1} = T_n \times 2$

Sequence 3: $T_1 = -5$ and $T_{n+1} = T_n + 10$

a State the first five terms.

b State whether the sequence is arithmetic, geometric or neither of these.

c State the sum of the first five terms.

d State the eighteenth term. (Use a calculator or spreadsheet.)

e State the sum of the first 18 terms. (Use a calculator or spreadsheet.)

12 Find the equation of the tangent to the curve $y = 2x^3 - x + 3$

a at the point (1, 4),

b at any points on the curve where the gradient is 23.

13 What do each of the following displays tell us about the rate of change of $f(x) = x^3 + 3x^2 + 4$?

a

Define f(x)=x³+3x²+4

$\frac{f(6)-f(1)}{5}$ 64

b

Define f(x)=x³+3x²+4

$\lim_{h\to 0}\left(\frac{f(5+h)-f(5)}{h}\right)$ 105

14 One of the graphs **A** to **D** shown below has $\frac{dy}{dx} = x(x + 3)$. Which one?

A

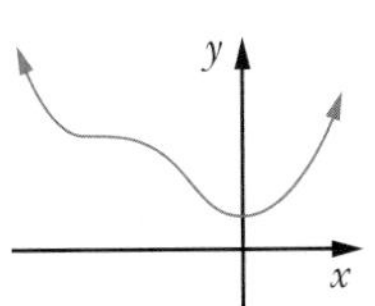

B

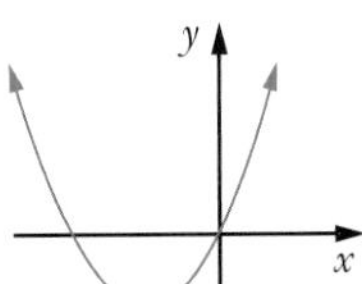

C

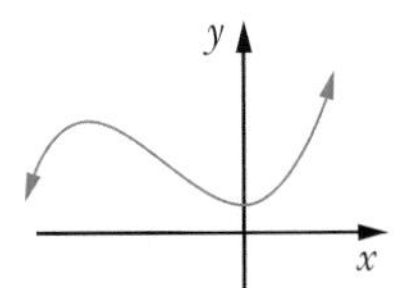

D

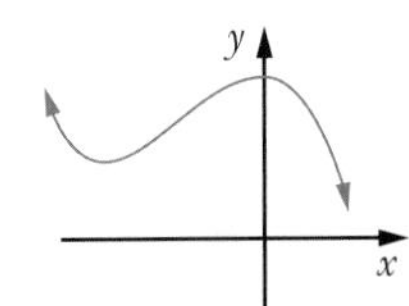

15 A curve is such that $\frac{dy}{dx} = x(x+6)(x-6)$.

a At how many places on the curve is the gradient zero?

b For $x \to \infty$ is the gradient positive or is it negative?

c For $x \to -\infty$ is the gradient positive or is it negative?

16 Figure 1 below shows a child's feeding bowl and figure 2 shows the same bowl with the shape of the interior shown.

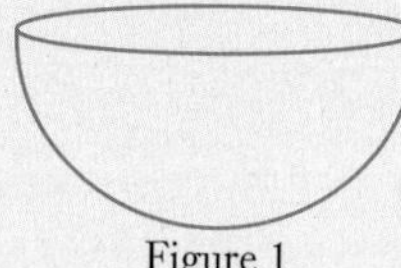

Figure 1

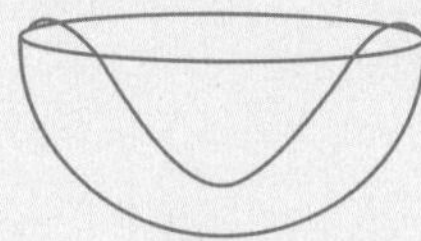

Figure 2

An unfortunate ant has found its way into the bowl and is at the bottom, hoping to get out. However the bowl's surface is very slippery so the ant may not be successful.

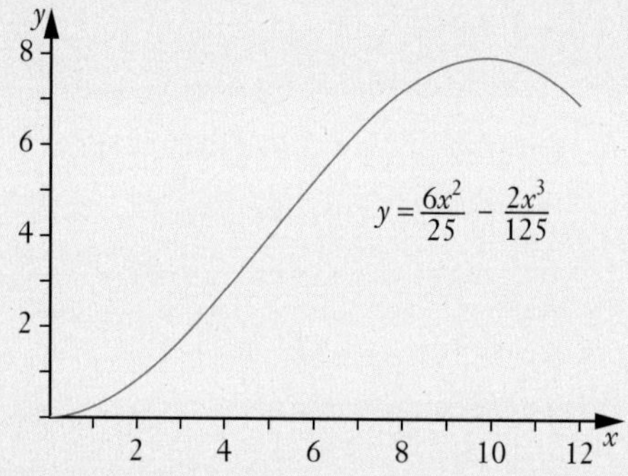

The above graph shows that the route the ant must follow, from the bottom of the bowl to the top, can be accurately modelled by part of the curve $y = \frac{6x^2}{25} - \frac{2x^3}{125}$.

The ant starts his (her?) climb to the top but, due to the slippery surface, will slip when the gradient of the slope is $\frac{144}{125}$.

Clearly showing the use of calculus and algebra, show that this gradient occurs twice in the section of the curve shown in the graph, stating the x-coordinate of each of these points and the y-coordinate of the *lower* point.

Check the x-coordinates just determined using a calculator that has the ability to differentiate functions.

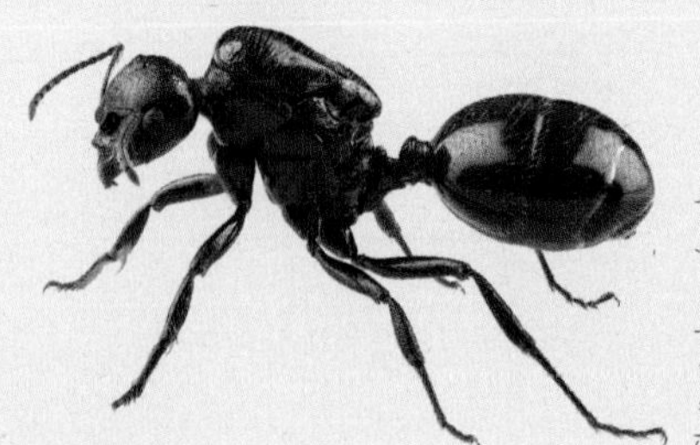

ISBN 9780170390408

Applications of differentiation

- Rates of change
- Using differentiation to locate stationary points of polynomial functions
- Global maximum and minimum values
- Applications
- Applications – extension
- Miscellaneous exercise six

Situation

Certain medical personnel in a country are concerned about the rate at which a particular disease is spreading through the population. The disease was first properly identified in 2003 and by the beginning of 2005 the number of people in the country known to have been suffering from the disease was 2050. Figures for later years suggest that the number of people in the country known to be suffering from the disease approximately follows the rule

$$N - t^3 + 5t^2 + 10t,$$

where N is the number of sufferers t years after the disease was first introduced to the country.

One of the country's top doctors is planning a speech about the disease and asks you to supply answers to the six questions given below.

Assuming the above equation and the 2005 figures are correct obtain answers to the questions.

- Is it true that the figures suggest that the disease first entered the country around the beginning of 1994?
- How many people in the country suffered from the disease at the beginning of 2000, even though it had not been properly identified at that time?
- How many people in the country suffered from the disease at the beginning of 2010?
- What was the average number of *new sufferers per year* in this ten year period?
- What was the rate of increase in the number of sufferers (in new sufferers per year) at the beginning of 2010?
- If nothing is done to alter the spread of the disease what is the rate of increase in the number of sufferers likely to be (in new sufferers per year) by the beginning of

  a the year 2040?

  b the year 2050?

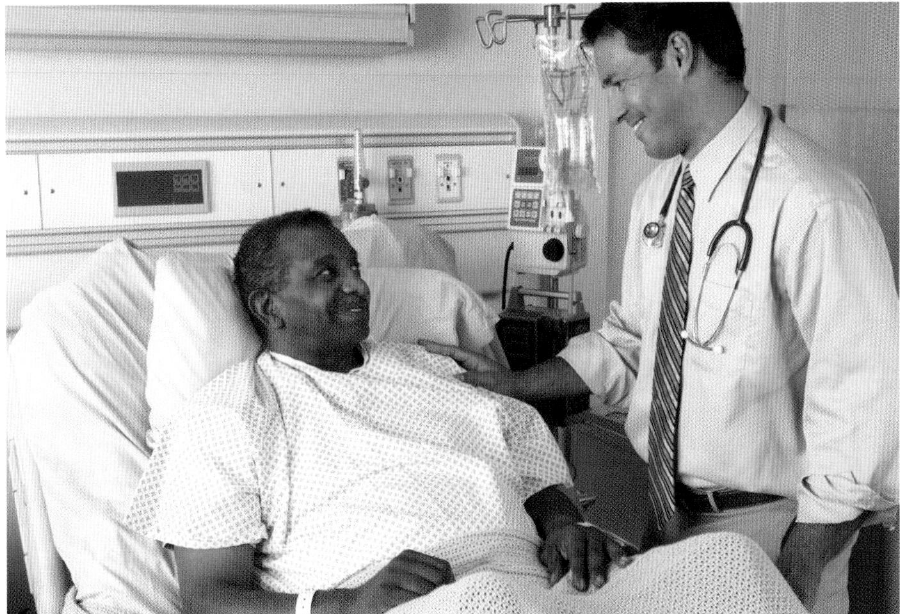

Shutterstock.com/Monkey Business Images

Instantaneous rates of change

Rates of change

Did you use differentiation in the situation on the previous page? The situation required a rate of change at a particular time to be determined. Differentiation gives a formula from which such a rate of change can be found.

In Chapter 5 we used differentiation to find $\frac{dy}{dx}$, the rate of change of y with respect to x. If other variables are involved, say volume, V, and time, t, then we can use differentiation to find $\frac{dV}{dt}$ the rate of change of volume with respect to time.

EXAMPLE 1

If $P = 5t^2 + 6t$ find an expression for the rate of change of P with respect to t.

Solution

If $P = 5t^2 + 6t$ then $\frac{dP}{dt} = 10t + 6$.

The rate of change of P with respect to t is given by $10t + 6$.

EXAMPLE 2

If $L = a^3 - 3a^2 + 5$ find the rate of change of L with respect to a when $a = 3$.

Solution

If $L = a^3 - 3a^2 + 5$ then $\frac{dL}{da} = 3a^2 - 6a$

If $a = 3$ then $\frac{dL}{da} = 9$

The rate of change of L with respect to a, when $a = 3$, is 9.

Alternatively, the same answers can be obtained from a calculator.

$\frac{d}{dt}(5t^2 + 6t)$ → $10 \cdot t + 6$

$\frac{d}{da}(a^3 - 3a^2 + 5) | a=3$ → 9

ISBN 9780170390408

EXAMPLE 3

The volume of a sphere is increasing in such a way that the volume, V cm^3, at time t seconds is given by: $V = 7500 + 3600t - 150t^2$ for $0 \le t \le 12$.

Calculate

a the volume when $t = 12$

b an expression for the rate of change of volume with respect to time

c the rate at which the volume is increasing (in cm^3/sec) when $t = 2$ and when $t = 10$.

Solution

a If $\quad V = 7500 + 3600t - 150t^2$

then $V(12) = 7500 + 3600(12) - 150(12)^2$

$= 29\,100$

When $t = 12$ the volume is 29 100 cm^3.

b If $\quad V = 7500 + 3600t - 150t^2$

then $\dfrac{dV}{dt} = 3600 - 300t.$

The instantaneous rate of change in the volume with respect to time is given by $3600 - 300t$.

Define v(t)=7500+3600t–150t^2
Done
v(12)
29100
$\frac{d}{dt}$(v(t))
–300·t+3600
$\frac{d}{dt}$(v(t)) | t=2
3000
$\frac{d}{dt}$(v(t)) | t=10
600

c Using $\dfrac{dV}{dt} = 3600 - 300t$:

For $t = 2$, volume is increasing at 3000 cm^3/sec.

For $t = 10$, volume is increasing at 600 cm^3/sec.

- As the above display suggests, answers can be obtained from a calculator without having to differentiate 'by hand'. Whilst you are encouraged to explore the capability of your calculator in this regard make sure you can use the appropriate calculus and algebraic methods yourself as well.

Exercise 6A

1 If $Q = 5r^2 + 3r - 4$ find an expression for the rate of change of Q with respect to r.

2 If $X = 3k + 3k^2 - 6k^3$ find an expression for the rate of change of X with respect to k.

3 If $T = 5r^3 - r^2 + 15r - 3$ find an expression for the rate of change of T with respect to r.

4 If $Q = 2p^4 + 3p^3 - 14p - 21$ find an expression for the rate of change of Q with respect to p.

5 If $P = (3t^2 - 2)(4t + 3)$ find an expression for the rate of change of P with respect to t.

6 If $A = 5t^2 + 6t - 1$ find the rate of change of A with respect to t when

a $t = 1$ **b** $t = 2$ **c** $t = 3$.

7 If $P = 3a^2 + 4$ find the rate of change of P with respect to a when

a $a = 2$ **b** $a = 3$ **c** $a = -4$

8 If $A = \pi r^2$ find the rate of change of A with respect to r when

a $r = 10$ **b** $r = 3$ **c** $r = \frac{70}{\pi}$

9 If $A = 2\pi r^2 + 20\pi r$ find, in terms of π, the rate of change of A with respect to r when

a $r = 3$ **b** $r = 7$ **c** $r = 10$

10 If $V = \frac{4}{3}\pi r^3$ find, in terms of π, the rate of change of V with respect to r when

a $r = 1$ **b** $r = 3$ **c** $r = 10$

11 A goldfish breaks the water surface of a pond when collecting food and causes a circular ripple to emanate outwards. The radius of the circle, in metres, is given by $r = \frac{2t}{5}$ where t is the time in seconds after the goldfish caused the ripple to commence.

a Find an expression for the area of the circle in terms of t.

b Find the area of the circle after two seconds.

c Find an expression for the rate at which the area is increasing with respect to t.

d Find the instantaneous rate of increase of A when $t = 3$.

12 A colony of bacteria is increasing in such a way that the number of bacteria present after t hours is given by N where $N = 120 + 500t + 10t^3$.

a Find the number of bacteria present initially (i.e. when $t = 0$).

b Find the number of bacteria present when $t = 5$.

c Find the average rate of increase, in bacteria/hour, in the first 5 hours.

d Find an expression for the instantaneous rate of change of N with respect to time.

e Find the rate the colony is increasing, in bacteria/hour, when

i $t = 2$ **ii** $t = 5$ **iii** $t = 10$

13 The total number of units, N, produced by a machinist, t hours into an 8 hour shift was found to approximately fit the mathematical model

$$N = 42t + 9t^2 - t^3 \text{ for } 0 \leq t \leq 8.$$

a How many units did the machinist produce in the eight hours?

b What was the machinist's average production rate, in units/hour, during the shift?

c How many units did the machinist produce in the final hour?

d Find the production rate, in units/hour, when

i $t = 1$ **ii** $t = 2$ **iii** $t = 3$

ISBN 9780170390408

14 A small crack in a water pipe allows water to escape from the pipe. The number of litres of water that has escaped, t minutes after the crack initially appeared is given by V, where $V = \frac{t}{1000}(t + 10)$.

a What volume of water leaked in

i the first ten minutes? **ii** the first twenty four hours?

b At what rate, in L/min, is the water leaking after

i 10 minutes? **ii** 2 hours? **iii** 24 hours?

15 A wildlife park is involved in a captive breeding program for an endangered species of deer. The program plans to release deer from the herd back into the wild as well as increasing the captive herd's population. The park starts with a captive population of forty deer and the breeding and release back into the wild will be such that the captive population P, t years later, approximately follows the mathematical rule:

$$P = 40 + \frac{t(t + 20)}{10}$$

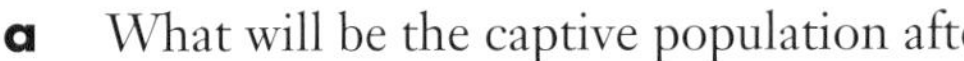

a What will be the captive population after

i 1 year **ii** 2 years **iii** 3 years **iv** 10 years?

b Find an expression for the rate of change of P with respect to t.

c Find the rate of change of P with respect to t (in deer/year) after

i 5 years **ii** 10 years **iii** 20 years

16 Following the survey of a particular mine, experts predict that with continued mining the quantity (T tonnes) of a particular ore remaining in the mine, t years after the survey was carried out would approximately fit the mathematical model

$$T = 20t^3 - 420t^2 - 8000t + 150\,000$$

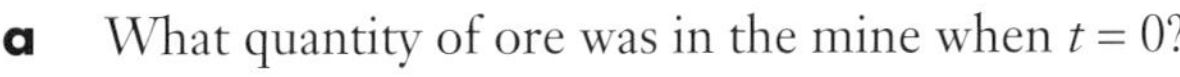

a What quantity of ore was in the mine when $t = 0$?

b What quantity of ore will be in the mine when $t = 10$?

c Find a rule for the rate of decrease of T in tonnes per year.

d Calculate the rate that T will be decreasing in tonnes per year when

i $t = 2$ **ii** $t = 4$ **iii** $t = 7$

17 An automatic puncture repair compound for bicycle tyres is being tested. The idea is that the compound can exist in vaporised form, mixed with the air in the tyre. When a puncture occurs the tyre pressure forces air out through the hole. This flow of air through the hole causes the glue-like compound to condense at the hole and repair the puncture. The compound is tested in a specially constructed balloon. The balloon is punctured and its volume is noted as it deflates. In the test the balloon had a volume V cm^3 where

$$V \approx 1000 - 4t + \frac{1}{10}t^2$$

with t being the time in seconds since the puncture occurred.

Determine

a the volume of the balloon when the puncture occurred

b the volume of the balloon two seconds after the puncture occurred

c an expression for the rate of change of V with respect to t

d the instantaneous rate of change of volume, in cm^3/sec, when

 i the puncture occurs

 ii $t = 3$

e how long it takes the compound to repair the puncture

The given formula for V can only make sense for $a \le t \le b$.

f Find a and b.

Stationary points

Sketching curves

Using differentiation to locate stationary points of polynomial functions

Note carefully the following points as the ideas are used in the examples that follow to locate the stationary points on the graphs of polynomial functions and to determine the nature of such points.

- At local maximum points, local minimum points and at points of horizontal inflection the gradient of the curve is momentarily zero.

  I.e. $\frac{dy}{dx}$ is zero at these points.

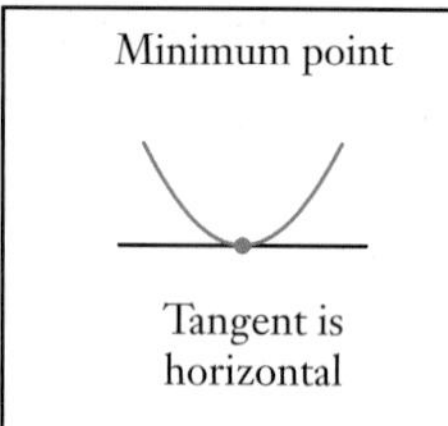

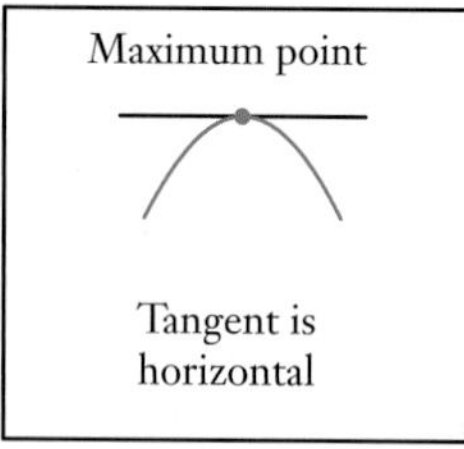

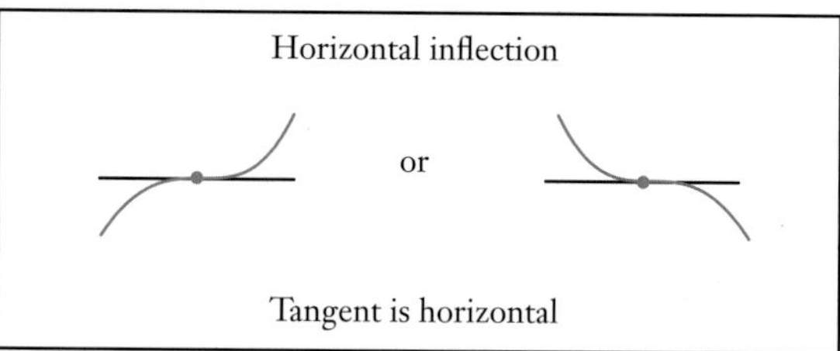

ISBN 9780170390408

- As we pass through a minimum turning point (in the direction of increasing x) the gradient changes from negative to positive.

  As we pass through a maximum turning point (in the direction of increasing x) the gradient changes from positive to negative.

  For horizontal inflection the gradient, though momentarily zero, does not change sign.

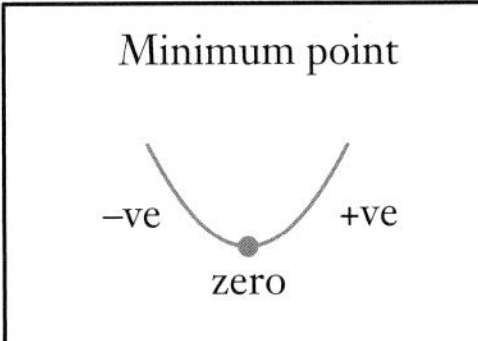

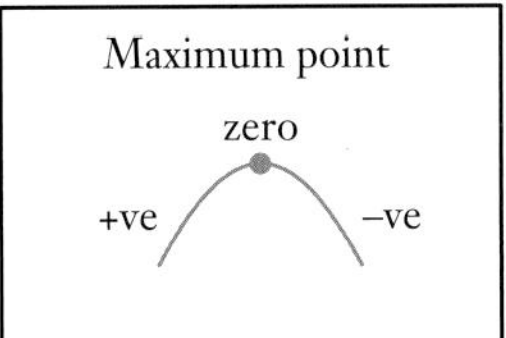

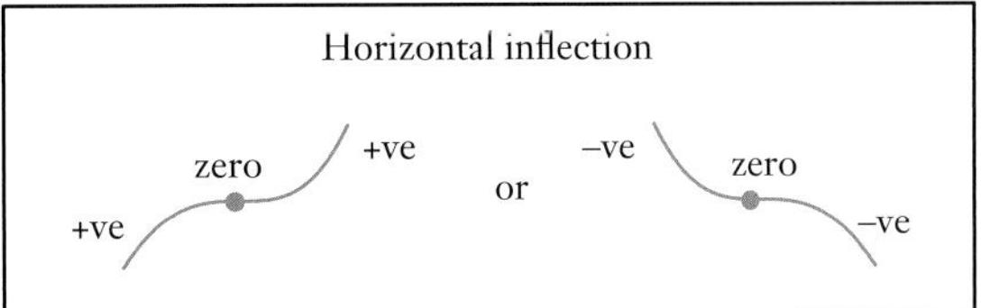

EXAMPLE 4

For the function $y = x^2 + 6x - 4$ use differentiation to determine the nature and location of any stationary points.

Solution

If $\quad y = x^2 + 6x - 4$

then $\quad \dfrac{dy}{dx} = 2x + 6$

At stationary points $\dfrac{dy}{dx} = 0 \qquad \therefore \quad 2x + 6 = 0$

i.e. $\quad x = -3.$

When $x = -3$,
$$\begin{aligned} y &= (-3)^2 + 6(-3) - 4 \\ &= 9 - 18 - 4 \\ &= -13. \end{aligned}$$

There is a stationary point at $(-3, -13)$.

Consider gradient either side of $x = -3$:

	$x = -3.1$	$x = -3$	$x = -2.9$
$2x + 6$	–ve	zero	+ve
	\	—	/

Thus $y = x^2 + 6x - 4$ has a minimum turning point at $(-3, -13)$.

Alternatively we could use our familiarity with the graphs of quadratic functions to state that the stationary point is a minimum because the coefficient of x^2 in the quadratic function is positive.

ISBN 9780170390408

EXAMPLE 5

For the function $y = 2x^3 - 6x^2$, and without the use of a calculator, determine

a the coordinates of any points where the graph of the function cuts the y-axis,

b the coordinates of any points where the graph of the function cuts (or touches) the x-axis,

c the behaviour of the function as $x \to \pm\infty$.

d the nature and location of any stationary points on the graph of the function.

Hence sketch the graph of the function.

Solution

a On the y-axis, $x = 0$.

If $x = 0$, $y = 2(0)^3 - 6(0)^2$

$= 0$.

The graph of the function cuts the y-axis at (0, 0).

b On the x-axis, $y = 0$.

If $y = 0$, $2x^3 - 6x^2 = 0$

i.e. $2x^2(x - 3) = 0$,

so $x = 0$ or 3.

The graph of the function cuts (or touches) the x-axis at (0, 0) and (3, 0).

c As x 'gets large' the x^3 term will dominate.

Thus as $x \to \infty$, $y \to \infty$ (and faster than x does),
and as $x \to -\infty$, $y \to -\infty$ (and faster than x does).

d If $y = 2x^3 - 6x^2$

then $\frac{dy}{dx} = 6x^2 - 12x$

$= 6x(x - 2)$.

At stationary points $\frac{dy}{dx} = 0$ $\quad\therefore\quad 6x(x - 2) = 0$

i.e. $x = 0$ or 2.

When $x = 0$, $y = 2(0)^3 - 6(0)^2$ and when $x = 2$, $y = 2(2)^3 - 6(2)^2$

$= 0$ $\qquad = -8$.

There are stationary points at (0, 0) and at (2, −8).

	Consider gradient either side of $x = 0$			Consider gradient either side of $x = 2$		
	$x = -0.1$	$x = 0$	$x = 0.1$	$x = 1.9$	$x = 2$	$x = 2.1$
$6x(x - 2)$	+ve	zero	−ve	−ve	zero	+ve
	/	—	\	\	—	/

Thus $y = 2x^3 - 6x^2$ has a maximum turning point at (0, 0) and a minimum turning point at (2, −8).

ISBN 9780170390408

The information from the previous parts of this question can be placed on a graph as shown below.

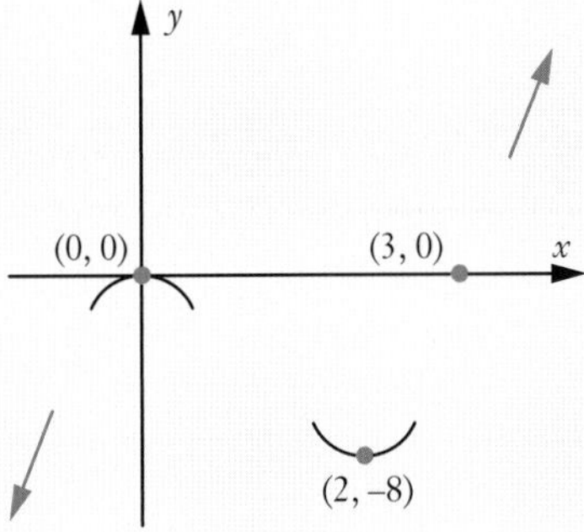

A sketch can be completed:

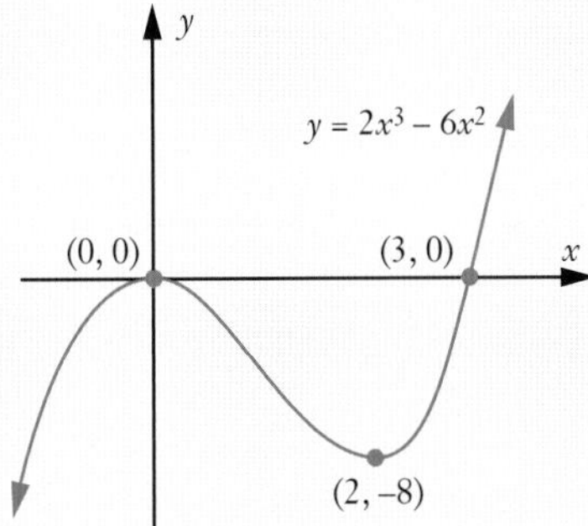

The reader should check the reasonableness of this sketch by viewing the graph of this function on a graphic calculator.

Note

- In the previous example the nature of each stationary point was determined by examining the sign of the gradient. We could have determined the nature of each point just from the shape that the sketch had to be to satisfy the behaviour as $x \to \pm\infty$ and the location of the turning points. From the diagram on the right (0, 0) must be a local maximum and (2, –8) must be a local minimum.

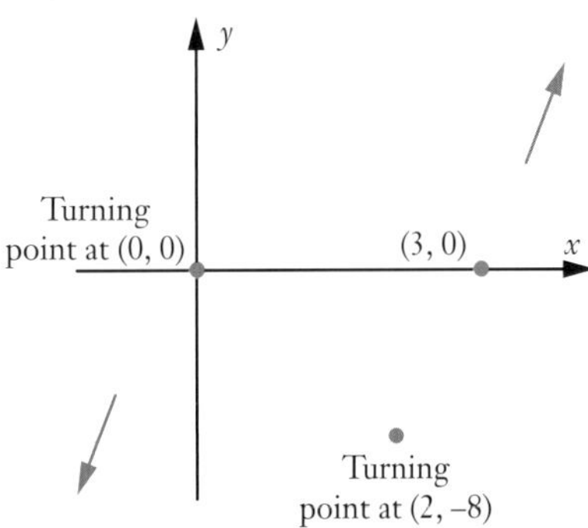

- Specific facilities on graphic calculators allow stationary points of polynomials, and other functions, to be readily located without the need to use differentiation. This ability is very useful but also needs care. In some cases the portion of the graph the display is showing may not be telling the whole story.

  For example the display below is for $y = x^2 - 0.01x^4$.

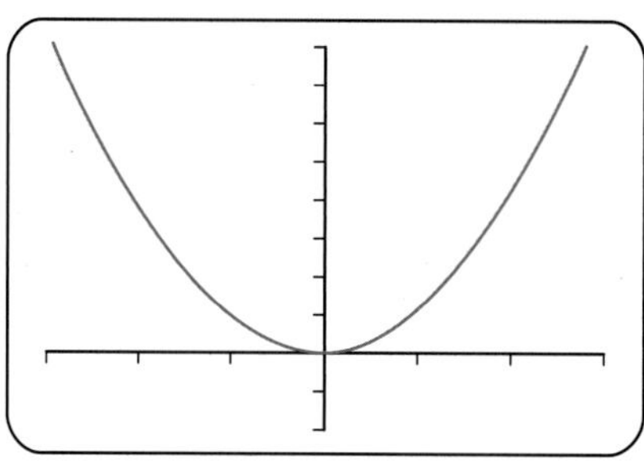

  At first glance it may appear to be similar to $y = x^2$ with a single turning point at (0, 0).

  However, as x gets large we would expect the $-0.01x^4$ term to dominate and would thus expect that as $x \to \pm\infty, y \to -\infty$.

  Sure enough, zooming out on the calculator we do indeed see that the original picture was not telling the whole story.

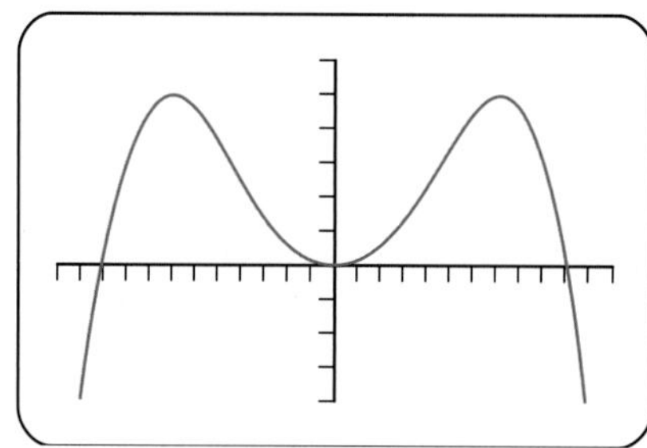

  The graph has three turning points, 2 maximums and one minimum. Using a calculus approach informs us where all of the stationary points are.

- If a question specifically requires that you do not use a calculator, or specifically requires you to show the use of differentiation (or **calculus**, of which differentiation is a part) and algebraic processes then proceed as follows:

  1 Differentiate y with respect to x to obtain $\frac{dy}{dx}$.

  2 Find the values of x for which $\frac{dy}{dx} = 0$.

  3 Find the values of y corresponding to each value of x from **2**.

  4 Either by considering the necessary shape of the graph, or by considering the sign of the gradient, determine the nature of the stationary points.

ISBN 9780170390408

Global maximum and minimum values

In some cases, we may be concerned with the maximum or minimum value a function can take for some interval $a \le x \le b$. We are then concerned with the global maximum (or minimum), which may or may not coincide with the local maximum (or minimum).

Greatest and least values

Application of optimisation

EXAMPLE 6

Using calculus and algebra determine the coordinates and nature of any stationary points on the graph of

$$f(x) = 9x^2 - x^3 - 15x + 11.$$

Hence determine the maximum value of $f(x)$ for

a $0 \le x \le 7$ **b** $-2 \le x \le 7$.

Solution

If $f(x) = 9x^2 - x^3 - 15x + 11$

then $f'(x) = 18x - 3x^2 - 15$

$= -3(x^2 - 6x + 5)$

$= -3(x - 1)(x - 5)$.

At stationary points $f'(x) = 0$, i.e. $-3(x - 1)(x - 5) = 0$

giving $x = 1$ or $x = 5$.

If $x = 1, f(x) = 4$. If $x = 5, f(x) = 36$.

Stationary points occur at (1, 4) and at (5, 36).

With $f(x) = 9x^2 - x^3 - 15x + 11$ the x^3 term will dominate for 'large x'.

Thus as $x \to +\infty$, $f(x) \to -\infty$

and as $x \to -\infty$ $f(x) \to +\infty$.

Thus (1, 4) is a local minimum point and (5, 36) is a local maximum point.

The global maximum will either occur at the local maximum or at the 'left end' of the function if the values of x allow us to go sufficiently far to the left.

a For $0 \le x \le 7$, $f(0) = 9(0)^2 - (0)^3 - 15(0) + 11$

$= 11$ which does not exceed $f(5)$.

Thus the global maximum for $0 \le x \le 7$ is 36.

b For $-2 \le x \le 7$, $f(-2) = 9(-2)^2 - (-2)^3 - 15(-2) + 11$

$= 85$ which does exceed $f(5)$.

Thus the global maximum for $-2 \le x \le 7$ is 85.

ISBN 9780170390408

Had we not been required to use calculus and algebra the global maximums determined in the previous example could have been obtained using a calculator that is able to determine the maximum value of a function for a given domain.

Again explore the capability of your calculator in this regard.

```
fMax(9·x^2–x^3–15·x+11, x, 0, 7)
                    (MaxValue=36, x=5)
fMax(9·x^2–x^3–15·x+11, x, –2, 7)
                    (MaxValue=85, x=–2)
```

Exercise 6B

1 A student used his old graphic calculator to locate the turning points on the curve

$$y = x^3 - 3x^2 - 45x - 20.$$

The display, see right, gave him the coordinates of the local minimum as (5.000 000 208 9, –195).

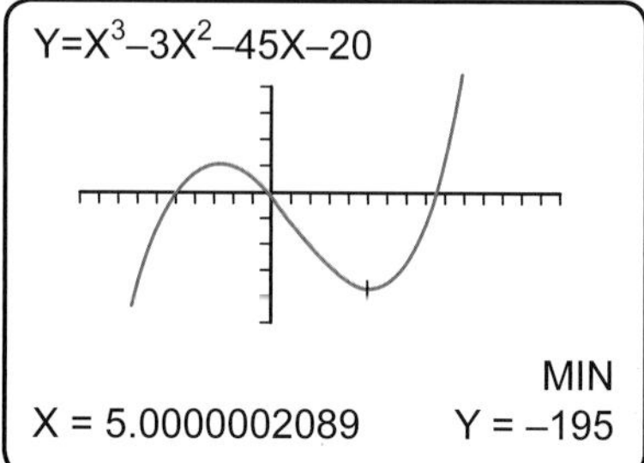

Use calculus to

a justify that the exact location of this minimum point is (5, –195),

b justify that the turning points displayed are the only ones the curve has,

c determine the exact coordinates of the local maximum point.

2 A student used her old graphic calculator to locate the turning points on the curve

$$y = x^3 + 1.5x^2 - 36x + 17.$$

The display, see right, gave her the coordinates of the local maximum as (–3.999 999 714 2, 121).

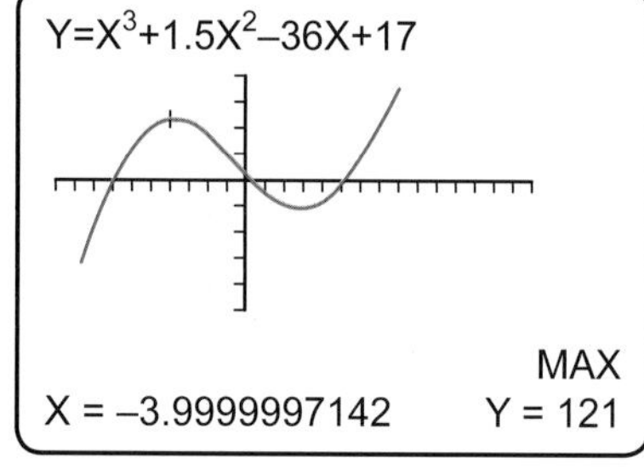

Use calculus to

a justify that the exact location of this maximum point is (–4, 121),

b justify that the turning points displayed are the only ones the curve has,

c determine the exact coordinates of the local minimum point.

Shutterstock.com/Brostock

ISBN 9780170390408

For questions **3** to **10**, and without the aid of a calculator:

- use calculus to determine the coordinates of any stationary points on the graph of the given function,
- indicate the nature of each stationary point,
- produce a sketch of the graph of the function including on your sketch the location of any stationary points, any points where the graph cuts the vertical axis and indicate the behaviour of the graph as $x \to \pm\infty$.

3 $y = x^3 + 3x^2 - 9x - 7$

4 $y = x^3 - 9x^2 + 15x + 30$

5 $y = 1 + 8x - 2x^2$

6 $y = x^5$

7 $y = x^4$

8 $y = 3x^2 - x^3$

9 $y = 2x^2 - 4x + 7$

10 $y = 3x^4 + 4x^3 - 12x^2 + 10$

11 For the function $y = x^3 + 6x^2 + 9x$, and without the use of a calculator, determine

a the coordinates of any points where the graph of the function cuts the y-axis,

b the coordinates of any points where the graph of the function cuts (or touches) the x-axis,

c the behaviour of the function as $x \to \pm\infty$,

d the nature and location of any stationary points on the graph of the function.

e Hence sketch the graph of the function.

f Determine the minimum and maximum value of y for $-5 \leq x \leq 1$.

12 Use calculus techniques to find the coordinates of any stationary points on the graph of $f(x) = 2x^3 - 3x^2$ and determine the nature of each.

Determine the minimum value of $f(x)$ for

a $x \geq 0$,

b $-1 \leq x \leq 5$.

Shutterstock.com/Doug Lemke

Gradient functions

Gradient functions

Applications

There are many occasions in real life when we look for the most desirable, most favourable or **optimum** situation. Finding the situation that involves maximum profit, most benefit, greatest growth, maximum effect, greatest comfort, greatest speed etc are all situations where the optimum situation involves a maximum. In other situations we might look for the least cost, the least effort, the lowest inflation, the minimum pain, the least discomfort, the slowest speed etc and in such cases are looking for a minimum level. Our ability to use calculus to determine the maximum or minimum values gives us a way of finding the optimum situations in various contexts.

EXAMPLE 7

What should be the dimensions of a rectangular shape of perimeter 20 cm if its area is to be a maximum?

Solution

We require maximum area. If the area is A cm^2 we need a formula A = ???.

Let the required rectangle have dimensions x cm by y cm.

y cm

x cm

Then $\quad A = xy. \quad [1]$

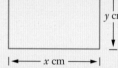

Now we cannot differentiate A because the right hand side of equation [1] involves two variables x and y.

However we do know that $2x + 2y = 20$

$\therefore \quad y = 10 - x. \quad [2]$

Substituting [2] into [1] gives $\quad A = 10x - x^2. \quad$ (A quadratic function.)

Thus $\quad \dfrac{dA}{dx} = 10 - 2x.$

If $\dfrac{dA}{dx} = 0$ then $\quad 0 = 10 - 2x \quad$ i.e. $\quad x = 5.$

From our knowledge of quadratic functions with a negative coefficient of x^2, or by examining the gradient of the function either side of $x = 5$, we know that $x = 5$ will give a local (and global) maximum.

Thus $x = 5$ gives a maximum value for A.

If $x = 5$ then, from equation [2], $y = 5$.

Thus for maximum area the rectangle should be a square of side 5 cm.

Points to note for solving optimisation problems:

- If a diagram is not given, then draw one if it helps.
- Identify the variable that is to be maximised, or minimised. If this variable is, say, C then you must find an equation with C as the subject. i.e. C = ???.
- If this equation for C involves two variables (other than C) find another equation that will allow you to substitute for one of the variables.

ISBN 9780170390408

- When you have C in terms of one variable, say x, then find the values of x for which $\frac{dC}{dx} = 0$.
- Use your knowledge of what the function must look like (perhaps it is a quadratic and therefore has one turning point … perhaps the behaviour as x 'gets large' helps … perhaps examine the gradient either side of the stationary point) to determine whether maximum or minimum.
- Check that the value of x for the required maximum, or minimum, is within the values that the situation allows x to lie and check that it gives the global maximum, or minimum.

EXAMPLE 8

The profit, $P, made by a company producing and marketing x items of a certain product is given by:

$$P = -x^3 + 30x^2 + 900x - 1000.$$

Clearly showing the use of calculus, find the value of x for maximum profit and determine this maximum profit.

Solution

We wish to maximise profit and we have a formula for P in terms of one variable, x. Thus we may differentiate.

$$\begin{aligned}\frac{dP}{dx} &= -3x^2 + 60x + 900\\ &= -3(x^2 - 20x - 300)\\ &= -3(x - 30)(x + 10).\end{aligned}$$

If $\frac{dP}{dx} = 0$, then $(x - 30)(x + 10) = 0$.

Solving gives $x = 30$ or -10 (-10 not applicable in this situation).

When $x = 30$, $P = -(30)^3 + 30(30)^2 + 900(30) - 1000$
$= 26\,000$.

If we also consider the y-axis intercept and $x \to \pm\infty$, a sketch can be made:

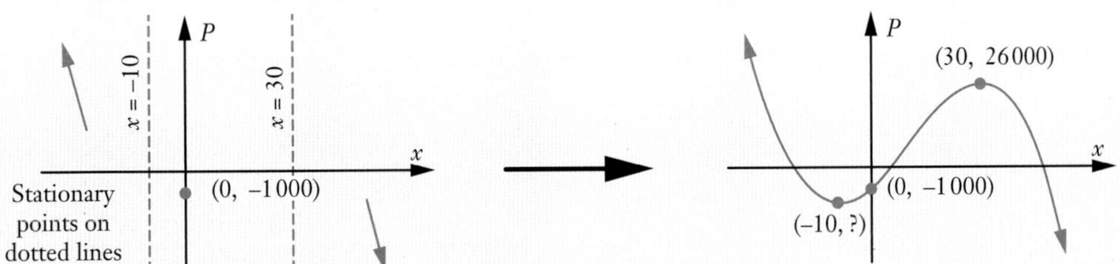

The sketch indicates that $x = 30$ will give the local *maximum* and for $x \geq 0$ this maximum will not be exceeded elsewhere.

For maximum profit the value of x should be 30 and the maximum profit would then be $26 000.

Note: In the previous example

- the nature of the stationary point at $x = 30$ can also be found by considering the gradient either side of $x = 30$:

	$x = 29$	$x = 30$	$x = 31$	
$-3(x - 30)(x + 10)$	+ve	zero	−ve	
	/	—	\	i.e. maximum point.

- viewing the graph of $y = -x^3 + 30x^2 + 900x - 1000$ on a calculator confirms the correctness of the sketch.

Exercise 6C

Use calculus to solve the following optimisation problems. (Use a calculator to assist with the arithmetic if you wish but clearly show the use of calculus to locate the optimum situation.)

1 If $X = t^3 - 15t^2 + 48t + 80$ find the value of t for which X has a local minimum value and find this minimum.

2 If $A = 60p + 12p^2 - p^3 - 500$ find the value of p for which A has a local maximum value and find this maximum.

3 If $A = xy$ and $x + 5y = 20$ find the maximum value of A and the values of x and y for which this maximum value occurs.

4 If $A = xy$ and $2x + 3y = 18$ find the maximum value of A and the values of x and y for which this maximum value occurs.

5 The total cost, $\$C$, and total revenue, $\$R$, arising from the production and marketing of x items of a certain product are given by

$$R = x(95 - x) \text{ and } C = 500 + 25x.$$

Given that Profit = Revenue – Cost, find the value of x that gives maximum profit and determine what this maximum profit will be.

6 The total cost for the production and marketing of x items of a certain product is $\$C$ where $C = 5000 + 60x$.

The revenue received from each item is $\$R$ where $R = 300 - x$.

Given that Profit = Revenue – Cost, find the value of x that gives maximum profit and determine what this maximum profit will be.

7 The organisers of a sheepdog competition have 100 metres of fencing available to fence an enclosure for some sheep. They wish to make the area rectangular and as large as possible. What dimensions should the enclosure have to maximise area if

a the 100 m of fencing is to be used for all four sides?

b an existing wall forms one side and the fencing is used for the other three?

ISBN 9780170390408

8 A manufacturer wishes to advertise a new product. He knows that advertising will increase sales but the advertising itself costs money. From past experience with a similar product the manufacturer expects that his profit \$$P$, after the advertising has been paid for, will be related to x, the number of thousands of dollars spent on advertising according to the rule

$$P = 50\,000 + 6000x - 100x^2.$$

How much should the manufacturer spend on advertising in order to maximise profit and what would this maximum profit be?

9 A rectangular box is to be made to the following requirements:

- The length must be one and a half times the width.
- The twelve edges must have a total length of 6 m.

Find the dimensions of the box that meets these requirements and that maximise the capacity.

10 An open cardboard box is to be made by cutting squares of side x cm from each corner of a square of card of side 60 cm and folding the resulting 'flaps' up to form the box. Find the value of x that will give the box a maximum capacity.

11 A long narrow sheet of metal, 8 metres by 24 cm, is to be made into a gutter by folding up equal widths of metal along each edge of the sheet to form the two identical vertical walls (see diagram).

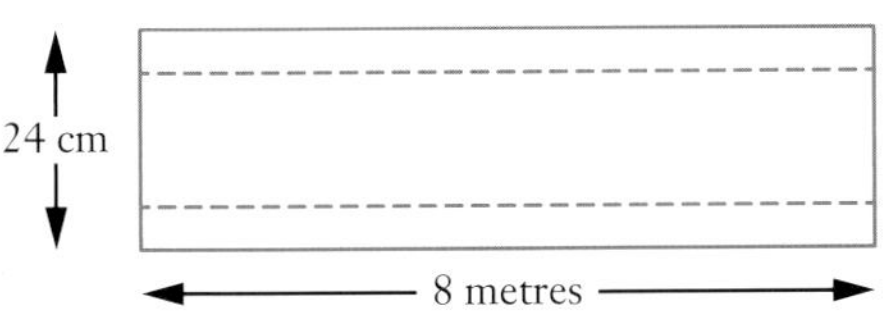

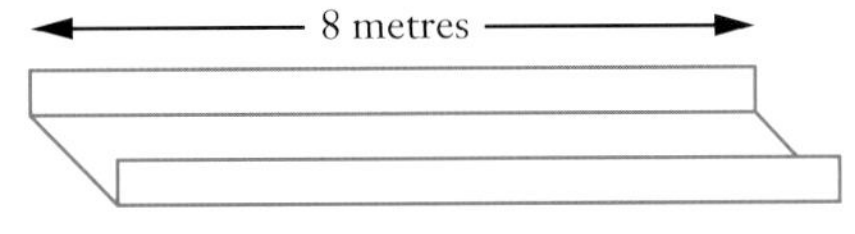

Use differentiation to determine how many centimetres should be turned up along each edge to maximise the capacity of the gutter for carrying water.

12 The organisers of a raffle are trying to decide the price they should charge for tickets. From past experience they feel confident that they can sell 7500 tickets if they charge \$1 per ticket. For each 10 cent rise in the price they estimate that they will sell 250 tickets less.

They need to raise \$5000 to cover the cost of prizes and printing.

If they set the price per ticket at \$$(1 + 0.1x)$, i.e. \$1 plus x lots of 10 cents, find

a an expression in terms of x for the profit the raffle will raise,

b the value of x for maximum profit.

For this maximum profit situation find

c the price of each ticket,

d the number of tickets they can expect to sell,

e the maximum profit.

13 A colony of bacteria is monitored in a laboratory over a 24 hour period ($0 \le t \le 24$) and its population, N, at time t was found to approximately follow the rule

$$N = 2t^3 - 57t^2 + 288t + 2900.$$

Determine the minimum and maximum value of N (nearest 100) for $0 \le t \le 24$.

14 A body is projected from an origin O and moves in a straight line such that its distance from O, t seconds after projection, is s metres where

$$s = \frac{t^3}{3} - 6t^2 + 50t, \ (t \ge 0).$$

The velocity of the body, v m/s, t seconds after projection is given by $\frac{ds}{dt}$.

- **a** How far is the body from the origin after three seconds?
- **b** Find an expression for the velocity of the body t seconds after projection.
- **c** What is the initial (i.e. $t = 0$) velocity of the body?
- **d** For what value of t ($t \ge 0$) is the body moving with minimum velocity and how far from O is the body at this time?

15 One small part of a construction project involves a metal rod spanning a 20 metre gap with the rod resting on supports at each end. The rod, which is not uniform, is expected to 'sag' somewhat under its own weight. The mathematical model of the situation predicts that this rod will take the shape of the curve

$$y = \frac{x}{50\,000}(20 - x)(x - 50), \text{ for } 0 \le x \le 20,$$

with x and y axes as shown in the diagram.

Clearly showing the use of calculus, but using your calculator to solve any equations that may result, determine the maximum sag in the rod (to the nearest mm) and where it occurs (as a distance from the origin, to the nearest centimetre).

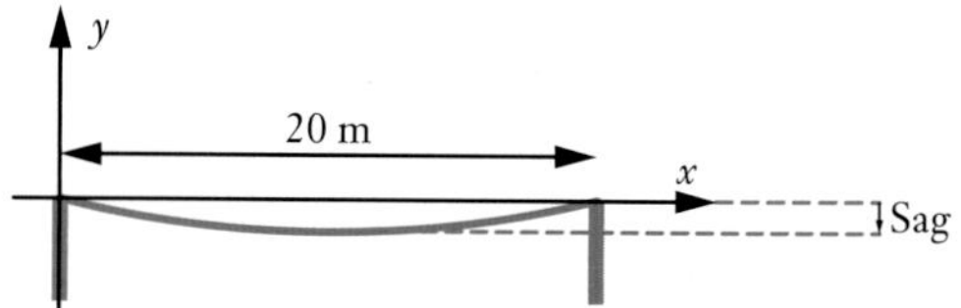

16 The owner of a large house decided to spend some money making it more secure. An expert analysed the situation and said that for \$5000 the security rating, R, of the property would rise from its current score of 30 points to 100 points. Every \$500 spent after that would lift the rating by 5 points.

The owner feels that the more security devices he pays for the more tedious he is going to find it to enter and leave his property with all that he will have to remember to lock/unlock, arm/disarm etc. He feels there is an owner convenience rating, C, which will go down by 2 points, from an initial 100 points, for every \$500 he spends over the \$5000 that he accepts is necessary.

As both the security rating and the convenience rating interest him he decides to multiply them together to form the 'secure but not inconvenient' rating Z.

Clearly showing your use of calculus, determine how much the owner should spend on security to maximise Z.

Shutterstock.com/Chakrapong Zyn

ISBN 9780170390408

Applications – extension

The next example, and the exercise that follows involve applications of differentiation for which the function involved contains negative or fractional powers. At the time of writing this text, the syllabus for this unit includes optimisation problems for polynomial functions only. Hence the optimisation questions in this next section should be regarded as an extension activity.

EXAMPLE 9

The rectangular block shown on the right has a square base of side x cm, a height of y cm and a volume of 80 cm^3. The base and top are to be covered with lacquer costing 5 cents/cm^2 and the sides with lacquer costing 4 cents/cm^2. Find the values of x and y for minimum cost.

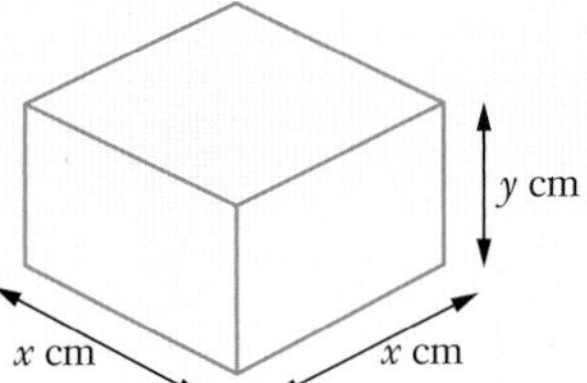

Solution

To minimise cost we need a formula. Cost = ???.

Cost of lacquering the base = $5x^2$ cents.

Cost of lacquering each side = $4xy$ cents.

Thus if the total cost is C cents then $\quad C = 10x^2 + 16xy \quad$ [1]

We cannot differentiate C at present because it involves *two* variables, x and y.

However we know that $\quad x^2y = 80$ (because the volume = 80 cm^3).

i.e. $\quad y = \dfrac{80}{x^2} \quad$ [2]

Substituting from [2] into [1] gives $\quad C = 10x^2 + \dfrac{1280}{x}$.

Thus $\quad \dfrac{dC}{dx} = 20x - \dfrac{1280}{x^2}$.

At stationary points $\dfrac{dC}{dx} = 0$, $\quad \therefore \quad 0 = 20x - \dfrac{1280}{x^2}$ giving $x = 4$.

From [2], if $x = 4$, $y = 5$ (and $C = 480$).

Given the context of the question x cannot be negative.

Considering $x \geq 0$:

With $C = 10x^2 + \dfrac{1280}{x}$ the y-axis will be an asymptote.

As $x \to 0$ (from the positive side), $y \to +\infty$.

As $x \to \infty$, $y \to +\infty$.

Thus (4, 5) must be a global minimum.

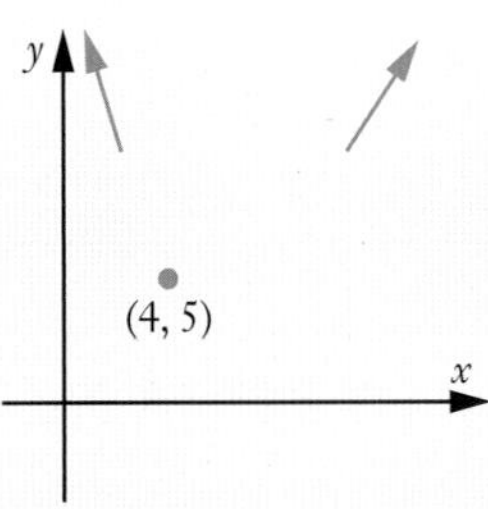

(Alternatively consider the sign of the gradient either side of $x = 4$.)

Thus to minimise cost the block should be made with a base 4 cm by 4 cm and a height of 5 cm, (i.e. $x = 4$ and $y = 5$).

Exercise 6D

1 If $P = 3r^2 + \frac{5}{r}$ find an expression for the rate of change of P with respect to r.

2 If $A = 400\sqrt{r}$ find the rate of change of A with respect to r when

a $r = 4$ **b** $r = 25$ **c** $r = 100$.

3 Use calculus to locate exactly the stationary points on the graph of $y = x + \frac{2}{x}$ and determine the nature of each by consulting a graphic calculator display.

4 Use calculus techniques to determine the exact coordinates of any stationary points on the curve

$$y = 5 - \frac{4}{x} - x$$

and, by considering the sign of the gradient on either side of any turning points determine whether maximum, minimum or horizontal inflection.

5 Use calculus techniques to determine the exact coordinates of any stationary points on the curve

$$y = 3x - \frac{96}{x^2}.$$

By considering the graph of the function for x close to zero and $x \rightarrow \pm\infty$ determine whether maximum, minimum or horizontal inflection.

6 The open rectangular box shown on the right is to have a square base of side x cm and a height y cm.

The volume of the box is to be 500 cm^3.

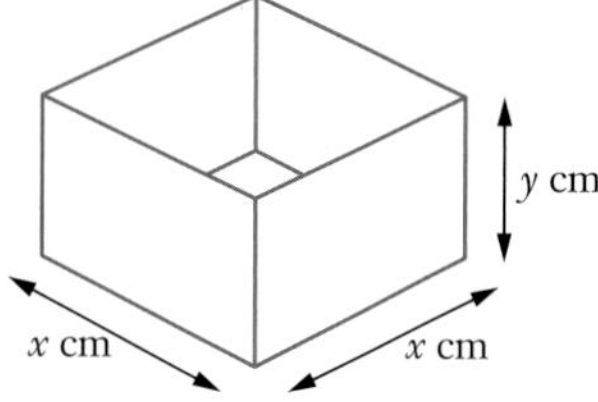

a Find an expression for y in terms of x.

b The box is to be made of card. Find, in terms of x, the area of card required to make each box, assuming no wastage.

c Find x and y for which this area is a minimum and then find this minimum area.

7 A food manufacturer wishes to package a product in cylindrical tins each of volume 535 cm^3. Find the base radius and height of tins that meet this volume requirement and that minimise the metal required to make them, i.e. minimum surface area. Give your answers in centimetres and correct to 1 decimal place.

8 A metal box company is asked to produce cylindrical metal tins, each with a volume of 535 cm^3. The base and top of each tin have to be made from thicker material than is used for the wall. This thicker material costs twice as much per cm^2 as the thinner material. Find, in centimetres and correct to one decimal place, the base radius and height of each tin for the cost of material to be a minimum.

ISBN 9780170390408

Miscellaneous exercise six

This miscellaneous exercise may include questions involving the work of this chapter, the work of any previous chapters, and the ideas mentioned in the Preliminary work section at the beginning of the book.

1 Express each of the following as a power of 5.

a 25 **b** 625 **c** 125

d 1 **e** $5 \times 5 \times 5$ **f** $5 \times 5 \times 5 \times 5 \times 5 \times 5$

g $(5 \times 5) \times (5 \times 5 \times 5)$ **h** $(5 \times 5 \times 5) \times 5$ **i** $(5 \times 5 \times 5)^2 \times 5$

j $\dfrac{5 \times 5 \times 5 \times 5 \times 5}{5 \times 5}$ **k** $\dfrac{5 \times 5 \times 5}{5}$ **l** $\dfrac{(5 \times 5 \times 5) \times (5 \times 5)}{5 \times 5 \times 5 \times 5}$

m $5^3 \times 5^7$ **n** $5^3 \times 5$ **o** $5^3 \times 5^7 \times 5^7$

p $5^5 \div 5^3$ **q** $5^8 \div 5^2$ **r** $5^{11} \div 5^8$

s $5^4 \times 5^3 \div 5^2$ **t** $5^3 \times 5^4 \div 5$ **u** $5^8 \div 5^3 \times 5^2$

v $5^5 \times 125$ **w** $5^5 \div 125$ **x** $5^8 \div (5^3 \times 5^2)$

y $3^2 + 4^2$ **z** $\dfrac{6^2 + 7 \times 2}{3 - 1}$

2 Without the assistance of a calculator, simplify each of the following, expressing your answers in terms of positive indices.

a $\dfrac{(a^3 \times \sqrt{a})^2}{a^3}$ **b** $\dfrac{(5b^{-2}a)^3}{25a^{-4}b^2}$ **c** $\dfrac{2^{n+1} + 2^{2n}}{2^n}$

d $\dfrac{5x^4 + 10x^7}{5x^3}$ **e** $\dfrac{2^x + 2^{x+3}}{9}$ **f** $\dfrac{3^{n+1} - 15}{5 \times 3^n - 25}$

3 An arithmetic progression is such that
- the fourth term is 130

and
- $T_{n+1} = T_n + 11$.

Determine the first six terms of the sequence.

4 A geometric progression is such that
- the fourth term is 2.8,

and
- $T_{n+1} = T_n \times 0.2$.

Determine the first five terms of the sequence.

5 Find **a** the average rate of change of the function $y = x^2 - 3x$ from the point P(3, 0) to the point Q(6, 18),

b the instantaneous rate of change of $y = x^2 - 3x$ at the point P(3, 0),

c the instantaneous rate of change of $y = x^2 - 3x$ at the point Q(6, 18).

ISBN 9780170390408

6 Find **a** the average rate of change of the function $y = x^3 - 3x$ from the point P(3, 18) to the point Q(6, 198),

b the instantaneous rate of change of $y = x^3 - 3x$ at the point P(3, 18),

c the instantaneous rate of change of $y = x^3 - 3x$ at the point Q(6, 198).

7 The diagram on the right shows a device for counting pills. The diagram shows the device containing 5 rows of pills.

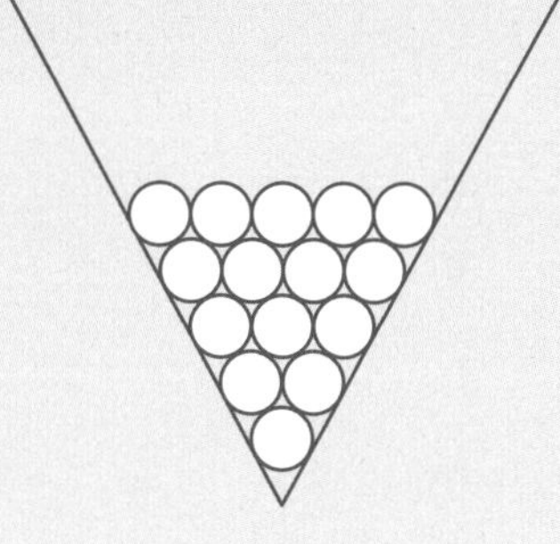

a How many pills are shown in the device?

b How many pills would be in such a device if it were to contain

i 10 complete rows?

ii 15 complete rows?

8 Evaluate the following sums:

a $3 + 12 + 21 + 30 + 39 + 48 + \ldots + 507$

b $S_{10} = T_1 + T_2 + T_3 + T_4 + \ldots + T_{10}$

$= 30 - 90 + 270 - 810 + \ldots - 590\,490$

c $6 + 12 + 24 + 48 + 96 + \ldots + 6\,291\,456$

d $100 + 80 + 64 + 51.2 + 40.96 + 32.768 + \ldots$

e $5 - 5 + 5 - 5 + 5 - 5 + 5 - 5 + 5 \ldots + 5$

9 The graph on the right shows a curve $y = f(x)$ with the tangents at $x = 1$ and at $x = 2$ drawn.

Use the tangents to determine the gradient of $y = f(x)$ at $x = 1$ and at $x = 2$.

As you may have realised, the graph is that of $y = x^3$. Check your previous answers using calculus.

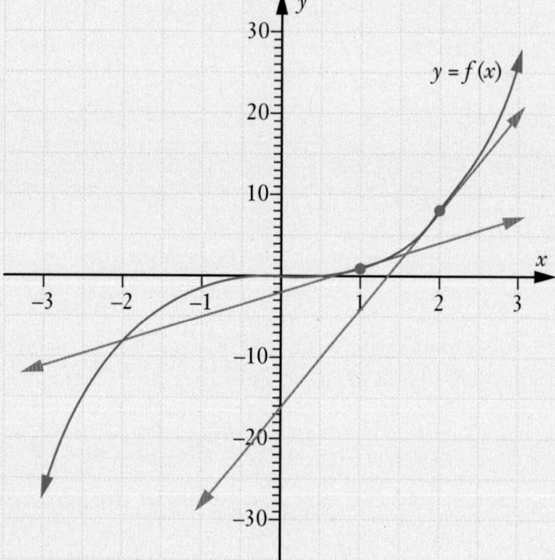

10 A curve is such that $\frac{dy}{dx} = (x + 4)(2x - 3)$.

At how many places on the curve is the gradient zero?

11 Use the formula:

$$\text{Gradient at P}(x, f(x)) = \lim_{h \to 0} \frac{f(x+h) - f(x)}{h}$$

to determine the instantaneous rate of change of

a the function $f(x) = x^2$ when $x = 5$.

b the function $f(x) = x^2 + x$ when $x = 1$.

c the function $f(x) = x^3 + x$ when $x = 2$.

ISBN 9780170390408

12 For the graph shown below state which of the points marked A → P are

a places on the curve where the function is zero, (4 points)

b places on the curve where the gradient is zero, (6 points)

c places on the curve where the gradient is positive, (4 points)

d places on the curve where the gradient is negative, (6 points).

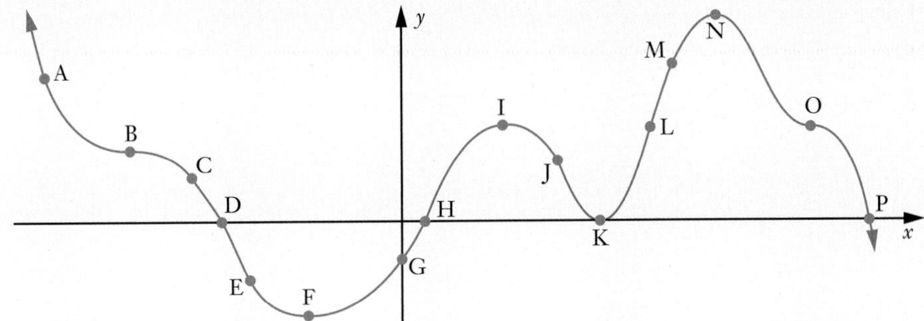

13 Where on the curve $y = x^2 + 5x - 4$ is the gradient the same as the gradient of the line with equation $y = 7x - 3$?

14 Find the coordinates of the point(s) on the following curves where the gradient is as stated.

a $y = 2x + \frac{1}{x}$. Gradient 1.

b $y = 3x - 4\sqrt{x}$. Gradient -1.

15 For $f(x) = 2x^4 - 5x^3 + x^2 - 2x + 6$ use your calculator to determine

a $f(21), f(31)$ and $f(41)$.

b $f'(21), f'(31)$ and $f'(41)$.

16 Clearly showing your use of differentiation and algebra find the equations of the tangents to the curve

$$y = x^3 + 3x^2 - 20x + 10$$

at any points on the curve where the gradient is equal to 25.

17 A butcher normally sells chicken fillets for \$10.50 per kg. During a week in which she has them on special for \$9.50 per kg she finds that her usual sales of 50 kg per week jumps to 70 kg per week.

a Assuming that the number of kg sold per week, N, obeys a rule of the form $N = ap + c$ where \$$p$ is the price per kg and a and c are constants, find a and c.

b Write an expression in terms of p for the total revenue the butcher receives for selling N kg at \$$p$ per kg.

c If the butcher pays \$7 per kg for the fillets write down an expression in terms of p for the profit she makes from buying and selling N kg.

d Find the value of p for maximum profit and, for this value of p, determine the number of kg sold and the profit.

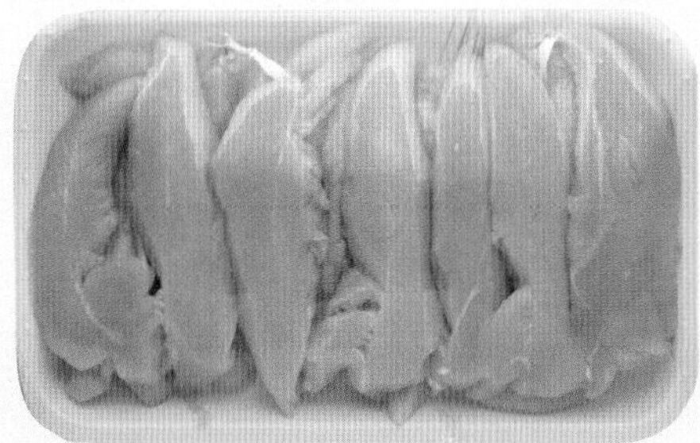

Shutterstock.com/MaraZe

18 Given that $p = 7 \times 10^{12}$ and $q = 2 \times 10^{11}$ evaluate each of the following, without the assistance of a calculator, giving your answers in standard form (scientific notation).

a $p \times q$ **b** $p + q$ **c** $p - q$

d $p \div q$ **e** $5pq$ **f** $\dfrac{p^2}{q}$

19 N, the number of organisms present in a certain culture of bacteria, t hours after observation commenced was found to approximately follow the rule

$$N = t^3 + 30t + 200.$$

Find

a the value of N when observation commenced,

b the value of N when $t = 10$,

c the average number of new organisms produced per hour during the first ten hours of observation,

d the instantaneous rate of change of N (in organisms per hour) when

i $t = 0$ **ii** $t = 5$ **iii** $t = 10$.

20 Explain what each of the following displays tell us about the rate of change of

$$f(x) = x^4 + x.$$

a

Define f(x)=x⁴+x

$\dfrac{f(3)-f(1)}{2}$ 41

b

Define f(x)=x⁴+x

$\lim_{h \to 0}\left(\dfrac{f(3+h)-f(3)}{h}\right)$ 109

21 The manufacturer of a certain fabric estimates that she can sell 500 m of the fabric each week if the price is $10 per metre. However, market research indicates that each 20 cents per metre price reduction will increase sales by 25 metres. If the manufacturer reduces the cost per metre by x lots of 20 cents find

a an expression for the cost per metre,

b an expression for the number of metres sold,

c an expression for the total revenue (income),

d the value of x that makes this total revenue a maximum and explain how you know that your value of x will give a *maximum*.

ISBN 9780170390408

Antidifferentiation

- Antidifferentiation
- Antidifferentiating powers of x
- Function from rate of change
- Can a calculator do the antidifferentiation for us?
- Miscellaneous exercise seven

Situation

A mathematics teacher taught this mathematics unit to two classes. One Tuesday morning she decided to give one of her classes a test involving ten functions to differentiate. She wrote the functions on the whiteboard and asked her students to write down the derivative of each one. She then went through each one, writing the answers on the board, and the students marked their work.

The teacher had her other class that afternoon and wanted to give them the same ten function test. She went to her classroom early to rub out the answers from the morning lesson. However, when she arrived at the classroom she found that someone had partly rubbed out the questions. What was left is shown below.

Tuesday — Test on differentiation.

$f(x)$		$f'(x)$
$5x^2$	1.	$10x$
$x^2 + 4$	2.	$12x$
$+ 3$	3.	7
	4.	4
$+ 3$	5.	$2x + 1$
	6.	x
	7.	$4x + 3$
	8.	$3x^2 - 5$
	9.	$12x^2 + 6x + 2$
	10.	$5x^4 - 1$

- Write a test involving the differentiation of ten functions that would be consistent with the information that was left on the board.
- Is the test you have written necessarily the same as that of others in your class?
- Is the test you have written necessarily the same as the one the teacher gave to her morning class?

Shutterstock.com/lightpoet

Antiderivatives

Antidifferentiation

Antidifferentiation is, as its name implies, the opposite of differentiation. If we differentiate x^2 we obtain the derivative, $2x$. If we antidifferentiate $2x$ we return to x^2, an **antiderivative** (or **primitive**) of $2x$. However we do have a problem: There are many functions that differentiate to $2x$.

If $y = x^2$ then $\frac{dy}{dx} = 2x$.

If $y = x^2 + 1$ then $\frac{dy}{dx} = 2x$.

If $y = x^2 - 1$ then $\frac{dy}{dx} = 2x$.

If $y = x^2 + 6$ then $\frac{dy}{dx} = 2x$. Etc.

Thus we say that the antiderivative of $2x$ is $x^2 + c$ where c is some constant. Given further information it may be possible to determine the value of this constant, as you will see in Example 2.

We can see the need for the '+ c' if we consider the situation graphically.

The diagram on the right shows the graph of $y = x^2$.

The gradient at $x = a$ will be the gradient of the tangent drawn at that point.

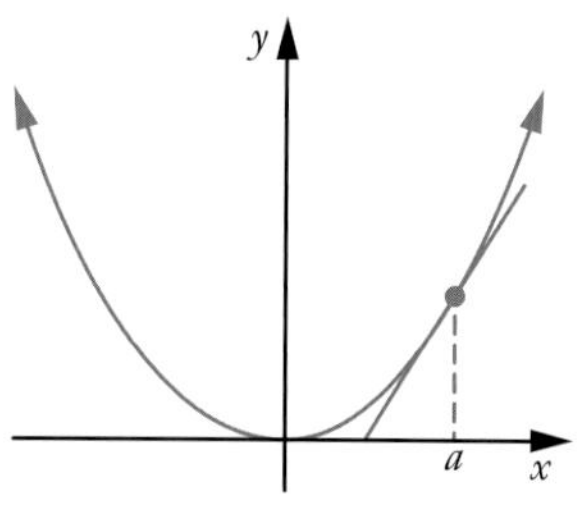

If we move $y = x^2$ up by c units the gradient at $x = a$ remains the same.

Thus, if $\frac{dy}{dx} = 2x$, antidifferentiation gives the 'family' of curves, $y = x^2 + c$.

Each member of this family has the gradient function $2x$ and each member can be obtained from any other by an appropriate vertical shift.

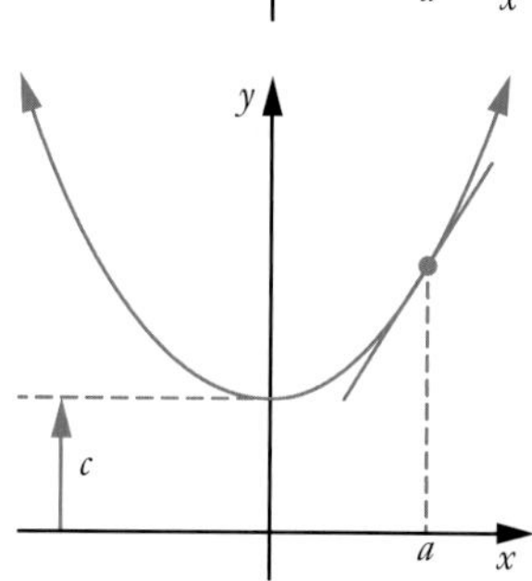

Antidifferentiating powers of x

In Chapter 5, we saw that to differentiate ax^n we used the rule 'multiply by the power and decrease the power by one'.

If $y = ax^n$ then $\frac{dy}{dx} = anx^{n-1}$.

To reverse this process we use the rule:

> 'Increase the power by one and divide by the new power.'

 ISBN 9780170390408

If $\frac{dy}{dx} = anx^{n-1}$ then $y = \frac{anx^n}{n}$ ← Power increased by one ← Divide by the new power

$= ax^n$

Thus if $\frac{dy}{dx} = ax^n$ then $y = \frac{ax^{n+1}}{n+1} + c$.

(Clearly the above rule cannot apply for $n = -1$. Such situations are beyond the scope of this unit.)

EXAMPLE 1

Find the antiderivative of each of the following.

a x^3

b $3x^4$

c 7

d $4x^3 + 12x^2 - 6x$

Solution

a If $\frac{dy}{dx} = x^3$,

then $y = \frac{x^4}{4} + c$

The antiderivative is $\frac{x^4}{4} + c$.

b If $\frac{dy}{dx} = 3x^4$,

then $y = \frac{3x^5}{5} + c$

The antiderivative is $\frac{3x^5}{5} + c$.

c If $\frac{dy}{dx} = 7$ (i.e. $7x^0$),

then $y = \frac{7x^1}{1} + c$

The antiderivative is $7x + c$.

d If $\frac{dy}{dx} = 4x^3 + 12x^2 - 6x$,

then $y = \frac{4x^4}{4} + \frac{12x^3}{3} - \frac{6x^2}{2} + c$

The antiderivative is $x^4 + 4x^3 - 3x^2 + c$.

The reader should confirm that differentiating each of the above answers does give the required gradient function.

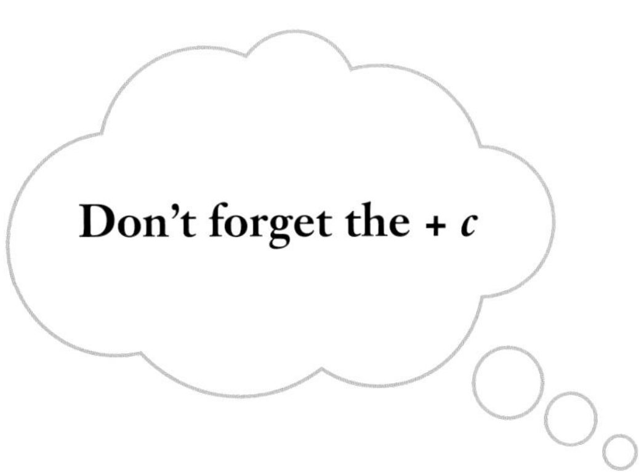

ISBN 9780170390408

EXAMPLE 2

If $\frac{dy}{dx} = 5 - 9x^2$ and, when $x = 1, y = 10$, find **a** y in terms of x,

b y, when $x = -1$.

Solution

a If $\frac{dy}{dx} = 5 - 9x^2$,

$$y = 5x - \frac{9x^3}{3} + c$$

$$= 5x - 3x^3 + c$$

We are told that when $x = 1, y = 10$.

Thus $10 = 5(1) - 3(1)^3 + c$

$10 = 5 - 3 + c$ giving $c = 8$.

$\therefore$ $y = 5x - 3x^3 + 8$

b If $x = -1$, $y = 5(-1) - 3(-1)^3 + 8$

$= 6$

When $x = -1, y = 6$.

Exercise 7A

Find the antiderivative of each of the following.

1 x^7 **2** x^5 **3** x^4

4 x^3 **5** x^2 **6** x

7 1 **8** $12x^2$ **9** $12x^5$

10 $8x^3$ **11** $14x$ **12** 6

13 $3x^2 + 6x$ **14** $6x^2 - 1$ **15** $7 + 12x^3$

16 $6x - 15x^4$ **17** $7 - 8x$ **18** $x^2 + 3$

19 $18x^5 + 1$ **20** $6x^2 + x$ **21** $12x^2 + 8x^3 + 2$

22 $3x^2 - 2x + x^5$ **23** $1 + x + x^2$ **24** $12x^3 + 6x + 5$

ISBN 9780170390408

For numbers **25** to **30** expand the given expression and then antidifferentiate.

25 $(3x + 4)(x + 2)$

26 $(9x - 1)(x + 1)$

27 $(x - 2)(x + 2)$

28 $(x + 1)(x - 3)$

29 $x^2(8x + 3)$

30 $4x(x^2 + 3x + 1)$

31 Find y in terms of x given that $\frac{dy}{dx} = 6x^2$ and $y = 5$ when $x = -1$.

32 Find y in terms of x given that $\frac{dy}{dx} = 3x + 2$ and $y = 0$ when $x = -2$.

33 Find y in terms of x given that $\frac{dy}{dx} = 3x^2 - 2x$ and $y = 6$ when $x = 1$.

34 Find y in terms of x given that $\frac{dy}{dx} = 6x^2 - 5$ and $y = 9$ when $x = 2$.

35 Find y in terms of x given that $\frac{dy}{dx} = 3 + 8x^3$ and $y = 6$ when $x = -1$.

36 If $f'(x) = \frac{3x^2}{2} + 4x - 1$ and $f(-2) = 4$ find

a $f(x)$,

b $f(2)$.

37 If $f'(x) = 3x - 6$ and $f(2) = 0$, find

a $f(x)$,

b $f(-2)$,

c a if $f(a) = 54$.

38 A curve has a gradient function of $2x + 7$ and passes through the points $(3, p)$ and $(-1, -9)$. Find the value of p.

39 A curve has a gradient function of $6x^2 - h$ and passes through the origin and the point $(4, 0)$. Find the coordinates of all the points where the curve cuts the x –axis.

40 A curve with a gradient function of $12x - 12$ cuts the x-axis at two points, $(3, 0)$ and $(k, 0)$. Find the value of k.

ISBN 9780170390408

Function from rate of change

In the previous chapter, *Applications of differentiation*, we applied our ability to differentiate to real life situations and obtained rates of change such as rate of change of profit, rate of change of area, rate of change of volume, etc. Antidifferentiation can be applied to such rates of change to return to the function for total profit, area, volume, etc.

EXAMPLE 3

The total revenue raised from the sale of x units of a particular product is $R(x)$ where $R(x)$ is such that:

$$\frac{dR}{dx} = \left(50 - \frac{x}{20}\right) \text{ dollars/unit.}$$

Given that the sale of zero items results in zero revenue find $R(x)$ in terms of x and determine the total revenue resulting from the sale of 100 items.

Solution

If $\frac{dR}{dx} = 50 - \frac{x}{20}$ then $R = 50x - \frac{x^2}{40} + c$.

iStock.com/jhjunker

We are told that when $x = 0$, $R = 0$.

i.e. $0 = 50(0) - \frac{(0)^2}{40} + c$ which gives $c = 0$.

$\therefore \quad R = 50x - \frac{x^2}{40}$ and $R(100) = 4750$.

Thus the total revenue function is $R = 50x - \frac{x^2}{40}$ and the revenue resulting from the sale of 100 items is \$4750.

Exercise 7B

1 If $\frac{dV}{dt} = 6t + 5$ find V in terms of t if $V = 30$ when $t = 0$.

2 If $\frac{dx}{dt} = 2t - 6$, and when $t = 2$, $x = -1$, find

a x in terms of t, **b** x, when $t = -2$, **c** t, when $x = 2$.

3 If $\frac{dA}{dr} = 4r + 12r^3$, and $A = 7$ when $r = 1$, find

a A in terms of r, **b** A, when $r = 2$.

ISBN 9780170390408

4 For each of the following, determine $C(x)$, the total cost function in dollars, from the given information.

a $\frac{dC}{dx} = (2x + 3)$ dollars per unit. $C(0) = 100$.

b $\frac{dC}{dx} = x(3x + 2)$ dollars per unit. $C(0) = 5000$.

5 For each of the following, determine $R(x)$, the total revenue function in dollars, given the following information.

a $\frac{dR}{dx} = 50$ dollars per unit. $R(0) = 0$

b $\frac{dR}{dx} = (50 - 0.05x)$ dollars per unit. $R(0) = 0$

6 If $R(x)$ is the total revenue from the sale of x items and is such that

$$\frac{dR}{dx} = (400 - 0.4x) \text{ dollars per unit}$$

and $R(0) = 0$, find the total revenue produced from the sale of 100 items.

7 A hole in a balloon causes it to deflate such that the rate of change of volume with respect to time is given by $\frac{dV}{dt} = -(20 + 10t)$ cm^3/s.

Find an expression for the volume of the balloon after t seconds given that when $t = 0$, $V = 7000$ cm^3.

8 A, the area of an oil slick, in m^2, t hours after observation commenced was found to be such that $\frac{dA}{dt} = 100$ m^2/h.

Is the slick increasing in area or decreasing?

If the area of the slick was 10 000 m^2 when observation commenced, find a formula for A in terms of t.

Getty Images/Kris Krug

9 \$$C$ is the total cost of producing x kg of a particular commodity. The rate of change of C with respect to x is \$40 per kg at all levels of production.

The fixed costs are \$1000, i.e. $C(0) = 1000$.

Find C as a function of x.

10 The total revenue for the production and sale of the x units of a particular commodity is given by \$$R(x)$ and is such that $\frac{dR}{dx} = \frac{2000 - x}{10}$.

Given that $R(0) = 0$, find $R(x)$ in terms of x and determine $R(1000)$.

ISBN 9780170390408

11 The manufacturers of a particular sports car find that their sales are usually about 20 cars per week. They run an advertising campaign lasting four weeks. During this time S, the total number of the sports cars sold since the model was first introduced changes such that

$$\frac{dS}{dt} = 20 + 20t - 3t^2$$

where t is the number of weeks the campaign has been running.

Find the number of cars sold **a** in the first week of the campaign,

b in the second week of the campaign,

c during the four week campaign.

Shutterstock.com/Neirfy

Can a calculator do antidifferentiation for us?

If you continue with later units of *Mathematics Methods* you will see that when we apply a particular 'summation' process to the function $f(x) = x^2$, the values we get fit the same function that antidifferentiating x^2 gives, i.e. $\frac{x^3}{3} + c$. This *summation* process is called *integration* and uses the following 'stretched S' symbol, $\int$, an s being used due to the link with a **S**ummation process.

Hence with this integration process, that you will meet in a later unit, giving the same function as the antidifferentiation process of this chapter, we tend to use the same stretched S symbol and the word integration when we are determining antiderivatives.

Thus:

- Instead of being asked to find the antiderivative of $6x^2 + 7$ we could be asked to **integrate** $6x^2 + 7$.
- The fact that the antiderivative of $6x^2 + 7$ is $2x^3 + 7x + c$
  could be written as $\int(6x^2 + 7)\,dx = 2x^3 + 7x + c$.
- The left-hand side of the previous equation tells us to integrate (or antidifferentiate) $6x^2 + 7$, 'with respect to x'. The stretched S and the dx act like a 'wrap' around the expression $6x^2 + 7$. The stretched S telling us that we are antidifferentiating, or integrating, and the dx telling us which variable is involved, in this case, x.

  In this way $\int(6t^2 + 7)\,dt = 2t^3 + 7t + c$.

ISBN 9780170390408

- Using this notation our general rule for antidifferentiating ax^n could be written:

$$\int ax^n\,dx = \frac{ax^{n+1}}{n+1} + c$$

- The '+ c', already encountered in this chapter, is called the **constant of integration** (or the constant of antidifferentiation). If sufficient information is given this constant can be determined, as has already been seen.
- Because integrals of the form $\int f(x)\,dx$ involve a constant of integration they are called **indefinite integrals**.

EXAMPLE 4

Find the following indefinite integrals.

a $\int 12x^2\,dx$,

b $\int (12x^2 + 2x - 3)\,dx$.

Solution

a $\int 12x^2\,dx = \frac{12x^3}{3} + c$

$= 4x^3 + c$

b $\int (12x^2 + 2x - 3)\,dx = \frac{12x^3}{3} + \frac{2x^2}{2} - 3x + c$

$= 4x^3 + x^2 - 3x + c$

So, to answer to the question posed earlier:

Can a calculator do antidifferentiation for us?

Some calculators do indeed have the ability to perform the antidifferentiation process but they tend to use the stretched S symbol when determining antiderivatives, as shown on the right.

$\int 6\cdot x^2+7\,dx$ $\quad 2\cdot x^3 + 7\cdot x$

$\int 12\cdot x^2\,dx$ $\quad 4\cdot x^3$

$\int 12\cdot x^2+2\cdot x-3\,dx$ $\quad 4\cdot x^3 + x^2 - 3\cdot x$

The displays may feature spaces for entries to be made above and below the integral sign, see right.

$\int_{\square}^{\square} 6\cdot x^2+7\,dx$ $\quad 2\cdot x^3 + 7\cdot x$

This is for *definite integrals*, a concept you will meet if you continue to higher units in this course of study. For this unit, if you have access to such a calculator simply leave such entries empty.

Note carefully: Calculators tend to omit the '+ c' so we must remember to include it with our answers when determining indefinite integrals.

ISBN 9780170390408

Exercise 7C

Find the following indefinite integrals.

1 $\int x^2\,dx$ **2** $\int x\,dx$ **3** $\int x^3\,dx$

4 $\int 2\,dx$ **5** $\int 10x^4\,dx$ **6** $\int 8x^3\,dx$

7 $\int (4x+1)\,dx$ **8** $\int (6x^2-5)\,dx$ **9** $\int (8x-7)\,dx$

10 $\int (x+9x^2)\,dx$ **11** $\int (x-1)\,dx$ **12** $\int (2x+3)(3x+1)\,dx$

13 $\int 6x(x+1)\,dx$ **14** $\int x^2(8x-3)\,dx$ **15** $\int 6x(x+1)^2\,dx$

Miscellaneous exercise seven

This miscellaneous exercise may include questions involving the work of this chapter, the work of any previous chapters, and the ideas mentioned in the Preliminary work section at the beginning of the book.

Express each of the following as a power of ten.

1 $10\,000$ **2** 0.1 **3** $10^9 \div 10^3$

4 $10^5 \times 1000$ **5** $10^5 \div 1000$ **6** $10^7 \times 10^5 \div 1\,000\,000$

7 $\sqrt{10}$ **8** $\sqrt[3]{10}$ **9** $\sqrt{1000}$

Find the value of n in each of the following.

10 $5^n = 1$ **11** $5^1 = n$ **12** $3^n = 243$

13 $3 \times 3^n = 243$ **14** $7^3 \times 7^n = 7^0$ **15** $10^6 \div 10^n - 100$

16 $10\,000 \times 10^n = 10^9$ **17** $2^n \div 4 = 2^6$ **18** $9^8 \div (9^4 \times 9^n) = 81$

19 A particular sequence is arithmetic with a common difference of 6 and a second term equal to 16. Define the sequence by stating the first term, T_1, and giving T_{n+1} in terms of T_n.
By how much does the sum of the first fifteen terms of this sequence exceed the fifteenth term?

20 Find $\frac{dy}{dx}$ for each of the following.

a $y = 5$ **b** $y = 5x + 5$ **c** $y = 5x^2 + 5x + 5$

d $y = 5x^3 + 5x^2 + 5x + 5$ **e** $y = x^2 + \sqrt{x}$ **f** $y = \frac{1}{2x^3}$

ISBN 9780170390408

21 If $f(x) = 4x^2 - 3x + 2$ find

a $f(3)$, **b** $f(-1)$, **c** $f'(x)$, **d** $f'(3)$.

22 Find the gradient of $y = x^2 + 2x + 1$ at the point (3, 16).

23 Find the gradient of $y = 3x^2 - 6x$ at the point (2, 0).

24 Use calculus and algebra to determine the coordinates of any point(s) on the curve $y = x^3 - 21x^2 - 96x + 2900$ where the gradient is zero.

25 Describe the graphs of each of the following in terms of a transformation of the graph of $y = 5^x$.

a $y = 5^x + 1$ **b** $y = 5^{x+1}$ **c** $y = 5^{-x}$ **d** $y = \frac{1}{5^x}$

26 Find the average rate of change of the function $y = x^2 + x$ from the point on the curve where $x = 3$ to the point on the curve where $x = 5$.

27 Find the coordinates of the points on the graphs of the given functions where the gradient is as stated.

a $y = 3x^2 + 5x$, gradient $= -1$. **b** $y = x^3 + 2x$, gradient $= 5$.

28 The quadratic function $y = (x - a)(x + b)$ cuts the x-axis at $(-3, 0)$ and $(7, 0)$. Given that $a > 0$ and $b > 0$, determine:

a the value of a and of b.

b the coordinates of the point where the function cuts the y-axis.

c the gradient of the quadratic at the points where it cuts the x-axis.

d the coordinates of any points on the quadratic where the gradient of the curve is equal to 6.

e the equation of the tangent to the quadratic at the point where the quadratic cuts the y-axis.

29 The curve $y = (x - a)(x + b)$, $a > 0$ and $b > 0$, cuts the x-axis at two points P and Q and cuts the y-axis at the point R(0, −12). If point P has coordinates (−4, 0) find a, b and the gradient of the curve at points P, Q and R.

30 The total cost, $C, of producing x units of a particular product is given by

$$C = 20\,000 + 2000x - 20x^2 + \frac{1}{15}x^3.$$

a Find an expression for the instantaneous rate of change of C with respect to x.

b Find the rate of change of C, with respect to x, when $x = 50$.

c If each unit sells for $2500 find an expression for $P(x)$, the profit made when x units of the product are produced and sold.

d Find the rate of change of P, with respect to x, when $x = 50$.

31 What can we conclude from the display on the right about the graph of

$$y = 4x^3 + 9x^2 - 210x + 75?$$

Define f(x) = $4x^3+9x^2-210x+75$
Done

Solve $\left(\frac{d}{dx}(f(x)) = 0, x\right)$
$x = -5$ or $x = \frac{7}{2}$

f(−5)
850

f(3·5)
−378.25

32 A rectangular box is to be made to the following requirements:

- The length, l cm, must be twice the width, w cm.
- The 12 edges must have a total length of 6 metres, i.e. $4l + 4w + 4h = 600$, where h cm is the height of the box.

a Copy and complete the following table:

Width (cm)	Length (cm)	Height (cm)	Volume (cm^3)
10			
20			
30			
40			

Continue your table for suitably chosen values for w in order to find, to the nearest centimetre, the dimensions of the box that meet the given requirements and that maximise the volume of the box.

b Express the volume of the box in terms of w and use calculus to confirm the answer you obtained in part **a**.

33 The diagram on the right shows the cross section of a tunnel with a truck just able to enter. With units in metres, and x- and y-axes as shown, the outline of the cross-section of the tunnel has equation

$$y = 12 - x^2.$$

Modelling the cross-section of this truck as a rectangle, with base and height as indicated, find the dimensions and area of such a rectangle that will just fit into the tunnel, if the area of the rectangle is to be a maximum.

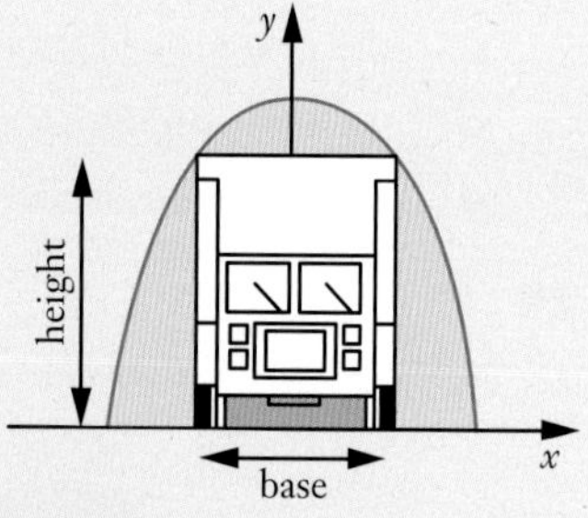

Shutterstock.com/zhangyang 13576997233

 ISBN 9780170390408

Rectilinear motion

- Displacement, speed and velocity
- Displacement from velocity
- Miscellaneous exercise eight

Situation

Two towns, A and B, are linked by straight road of length 60 km. The distance time graph shown below is for the motion of a cyclist travelling from town A to town B, and a delivery truck making the round trip from A to B and back to A again.

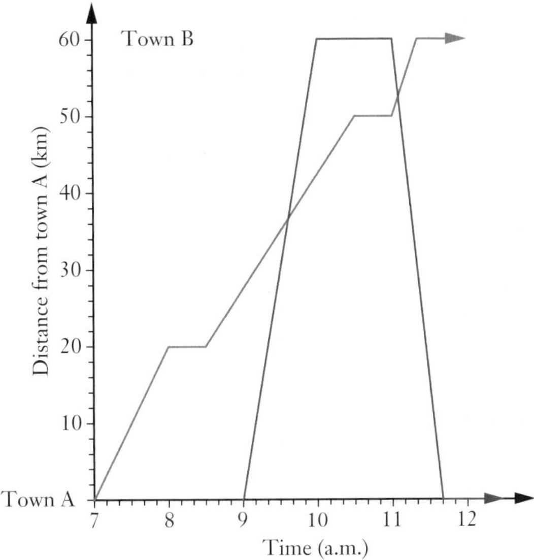

- When did the cyclist leave town A?
- When did the cyclist reach town B?
- The cyclist stopped twice for a rest. How long was each stop?
- What speed did the cyclist maintain
  - **a** prior to the first stop?
  - **b** between the two stops?
  - **c** after the second stop?
- What speed did the delivery truck maintain
  - **a** from town A to B?
  - **b** from town B back to A?
- Estimate the time and distance from A of the place where the delivery truck passed the cyclist when they were both travelling towards B.
- Estimate the time and distance from A of the place where the delivery truck passed the cyclist when the truck was returning to A.

Shutterstock.com/Milos Muller

ISBN 9780170350495

The situation on the previous page involved a position-time graph, or travel graph. As you probably realised, the speed of the cyclist or the delivery truck could be found from the gradient of the appropriate line. Horizontal lines indicated stops, steeper lines indicated greater speed etc. The graph involved only straight lines. If curves were involved the speed could again be found from the gradient but we would then need to draw tangents or use differentiation to determine instantaneous rates of change.

Notice here that we are talking about the *speed* of something – what is the difference between speed and velocity?

Displacement, speed and velocity

Rates of change

Graphs of rates of change

Straight-line motion 1

Straight-line motion 2

One of the commonest rates of change that concerns us is the rate at which we change our location. If we measure our location as a **displacement** from some fixed point or origin, then the rate at which we change our displacement is our **velocity**. Displacement and velocity are what are called **vector** quantities, they have magnitude (size) *and* direction. For example, our displacement might be 5 kilometres north of some origin and we might be travelling with a velocity of 6 km/hour south. On the other hand distance and speed are **scalar** quantities. They have magnitude only. For example we might travel a distance of 6 km at a speed of 60 km/hour. This chapter considers only rectilinear motion – **motion in a straight line**. For motion in a straight line there are only two possible directions and these are distinguished by the use of positive and negative.

Consider the following diagram which shows the location and velocity of four objects (referred to as 'bodies', but by that we do not mean cadavers!).

The origin, O, is as shown and positive x is to the right.

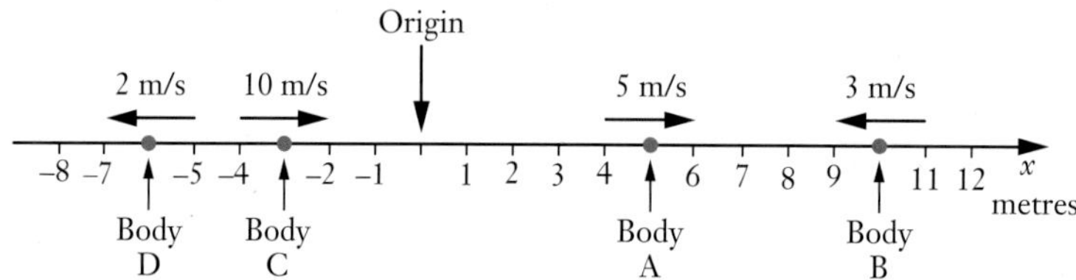

The table below shows how the directions of the displacement and velocity vectors can be indicated by use of positive and negative.

Body	Distance from O	Displacement from O	Speed	Velocity
A	5 metres	5 metres	5 m/s	5 m/s
B	10 metres	10 metres	3 m/s	–3 m/s
C	3 metres	–3 metres	10 m/s	10 m/s
D	6 metres	–6 metres	2 m/s	–2 m/s

ISBN 9780170350495

If our displacement, x metres, from some fixed origin, O, is a function of time then

$$\frac{dx}{dt}$$

gives the rate of change of this displacement with respect to time, i.e. it gives our velocity, v m/s.

For example, if $\quad x = 5t^3 + 6t^2 + 7t + 1$

then $\quad v = \frac{dx}{dt} = 15t^2 + 12t + 7.$

Similarly, if $\quad x = 5t^2 + 6 \quad$ then $\quad v = 10t;$

if $\quad x = 5t^3 - 3t + 1 \quad$ then $\quad v = 15t^2 - 3.$

EXAMPLE 1

A body moves in a straight line such that its displacement from an origin O, at time t seconds, is x metres where $x = t^3 + 6t + 5$.

Find the displacement and velocity when $t = 3$.

Solution

$x = t^3 + 6t + 5.$

When $t = 3$, $\quad x = (3)^3 + 6(3) + 5$

$= 50.$

$v = \frac{dx}{dt} = 3t^2 + 6.$

When $t = 3$, $\quad v = 3(3)^2 + 6$

$= 33.$

When $t = 3$ the displacement is 50 m, and the velocity is 33 m/s.

Or, using a calculator:

$t^3+6t+5 \mid t = 3$ 50

$\frac{d}{dt}(t^3+6t+5) \mid t = 3$ 33

ISBN 9780170350495

EXAMPLE 2

A body moves in a straight line such that its displacement from an origin O, at time t seconds is x metres where $x = 5t^2 + 7t + 3$.

Find **a** the displacement from O when $t = 0$,

b the initial (i.e. $t = 0$) velocity of the body,

c the value of t for which the velocity is 52 m/s.

Solution

Shutterstock.com/Scanrail1

a When $t = 0$, $x = 5(0)^2 + 7(0) + 3 = 3$

The displacement from O when $t = 0$ is 3 m.

b If $x = 5t^2 + 7t + 3$ then $v = \frac{dx}{dt}$

$= 10t + 7.$

Thus when $t = 0$, $\quad v = 7.$

The initial velocity of the body is 7 m/s.

c If $v = 10t + 7$ then for $v = 52$ m/s we have $52 = 10t + 7$

i.e. $45 = 10t$

so $\quad t = 4.5.$

The body has a velocity of 52 m/s when $t = 4.5$.

EXAMPLE 3

A particle is initially at an origin O. It is projected away from O and moves in a straight line such that its displacement from O, t seconds later, is x metres where $x = t(12 - t)$.

Find **a** the speed of initial projection,

b the distance the particle is from O when $t = 3$ and when $t = 7$,

c the value of t when the particle comes to rest and the distance from the origin at that time,

d the distance the particle travels from $t = 3$ to $t = 7$.

Solution

a If $x = 12t - t^2$ then $\quad v = \frac{dx}{dt}$

$= 12 - 2t.$

Thus when $t = 0$, $\quad v = 12.$

The speed of projection is 12 m/s.

ISBN 9780170350495

b If $t = 3$ then $x = 3(12 - 3)$

$= 27$

If $t = 7$ then $x = 7(12 - 7)$

$= 35$

The particle is 27 m from O when $t = 3$ and 35 m from O when $t = 7$.

c With $v = 12 - 2t$ then the particle being at rest means $12 - 2t = 0$,

i.e. $t = 6$.

When $t = 6$, $x = 6(12 - 6)$

i.e. $x = 36$.

The particle is at rest when $t = 6$ and it is then 36 m from O.

d When $t = 3$, the particle is 27 m from O.

When $t = 7$, the particle is 35 m from O.

However, the distance travelled in this time is not simply $(35 - 27)$ m. Our answer to **c** indicates that the particle stopped when $t = 6$,

i.e. at $x = 36$ m.

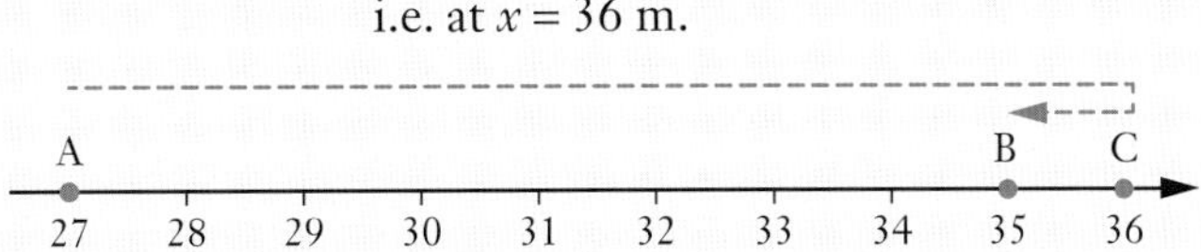

From $t = 3$ to $t = 7$ the particle travels from A to C to B, i.e. 9 m + 1 m = 10 m.

The particle travels 10 m from $t = 3$ to $t = 7$.

Exercise 8A

1 A long straight road links three towns A, B and C with B between A and C. From town A it is 130 km to B and a further 140 km to C. A truck leaves A at 8 a.m. and travels to B. For the first half hour the truck maintains a steady speed of just 60 km/h due to speed restrictions. After this the truck is able to maintain a higher speed and arrives in town B at 9.30 a.m. Unloading and loading in town B takes one hour and then the truck travels on to C maintaining a steady 80 km/h for this part of the journey.

A car leaves A at 9 a.m. that same morning and travels directly to C. Subject to the same speed restrictions it too maintains a steady 60 km/h for the first half hour. After this first half hour the car then maintains a steady 100 km/h all the way to town C.

Draw a distance time graph for this situation and use your graph to answer the following questions.

a When does each vehicle reach town C?

b What steady speed did the truck maintain from 8.30 a.m. to 9.30 a.m.?

c What was the average speed of the truck from A to B (to the nearest km/h)?

d When and where did the car pass the truck?

2 Copy and complete the given table for the situation shown below.

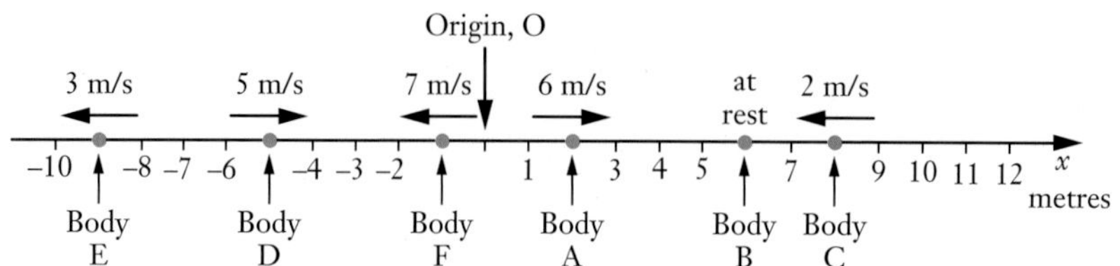

Body	Distance from O	Displacement from O	Speed	Velocity
A				
B				
C				
D				
E				
F				

3 For each of the following, x metres is the displacement of a body from an origin O, at time t seconds. Find the instantaneous velocity of the body for the given value of t.

a $x = t^2, t = 2.$ **b** $x = 2t^3 + 7t - 1, t = 2.$

For questions **4** to **9**, x metres is the displacement of a body from an origin O at time t seconds.

For each question find **a** the initial displacement from O,

b the initial velocity,

c the speed of the body when $t = 3$.

4 $x = 3t^2 + 5t + 6$ **5** $x = t(t - 3)$ **6** $x = 2t^3 - 3t^2 + t$

7 $x = 6t + 3$ **8** $x = 2t^2 - 20t - 3$ **9** $x = 1 - 6t + 12t^2 - 8t^3$

10 The displacement of a body from an origin O, at time t seconds is x metres where

$$x = 3t^2 + 2t + 1.$$

Find the displacement and velocity and of the body when $t = 2$.

11 The displacement of a body from an origin O, at time t seconds is x metres where

$$x = 2t^3 - 3t^2 + 4t - 1.$$

Find the displacement and velocity of the body when $t = 1$.

For questions **12** to **14**, x metres is the displacement of a body from an origin at time t seconds.

For each question find **a** the displacement when $t = 1$,

b the velocity when t = 1,

c t, (≥ 0), when the velocity is 20 m/s.

12 $x = t^2 + 6t + 1$ **13** $x = t^3 - 4t^2 + 4t + 5$ **14** $x = t^3 - 3t^2 + 11t + 3$

ISBN 9780170350495

15 The displacement of a body from an origin O, at time t seconds, is x metres where

$$x = 27t + 3t^2 - \frac{t^3}{3} - 90, t \geq 0.$$

Find the displacement of the body from O when the velocity is zero.

For questions **16** to **19**, x metres is the displacement of a body from an origin at time t secs.

For each question find

a the displacement when $t = 2$,

b the displacement when $t = 8$,

c t, (≥ 0), for which the body is at rest,

d the distance travelled from $t = 2$ to $t = 8$,

e the distance travelled in the 3rd second.

16 $x = 20t - t^2 + 9$

17 $x = t^2 - 8t + 20$

18 $x = 12t - t^2 + 20$

19 $x = t^3 - 6t^2 - 15t + 140$

20 A body is projected from ground level, vertically upwards, with an initial speed of 50 m/s. The body attains a height of x metres above ground level, t seconds after projection where $x = 5t(p - t) + q$ with p and q constant.

a Find p and q.

b Find the height and velocity of the body four seconds after projection.

c What will be the speed of the body when it reaches its highest point and how high will it be then?

21 A body is projected vertically upwards from ground level, with initial speed b m/s. The height it attains, t seconds after projection is x metres where x is given by

$$x = bt - 5t^2.$$

Find the speed of projection if the body just reaches a height of 180 m.

22 Two particles, A and B, are travelling along the same straight line. Their displacements from an origin O, at time t seconds, are x_A and x_B where

$$x_A = 2t^2 + 3t - 6 \qquad \text{and} \qquad x_B = 30 + 60t - 3t^2.$$

These rules apply for $0 \leq t \leq t_1$ where t_1 is when the particles collide.

a Find t_1.

b Find how far from O the collision occurs.

c Which of the following occur when $t = t_1$?

- A catches up with B,
- B catches up with A,
- A and B collide 'head-on'.

Shutterstock.com/nikkytok

ISBN 9780170350495

Displacement from velocity

From our understanding of antidifferentiation, or integration, as the reverse of differentiation it follows that integrating velocity with respect to time will give displacement.

i.e. $$x = \int v\,dt$$

Thus:

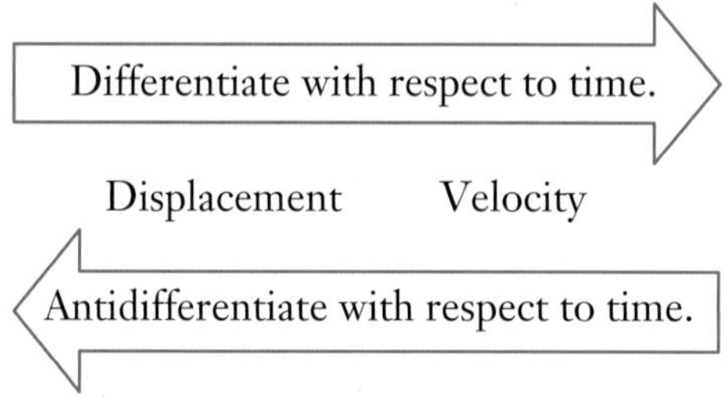

Remember that antidifferentiation, or integration, will introduce a constant which, given sufficient information, may be determined.

EXAMPLE 4

A particle travels along a straight line with its velocity at time t seconds given by v m/sec where $v = 3t^2 + 2$.

The initial displacement of the particle from a point O on the line is ten metres. Find the displacement from O when t = 5.

Solution

If $v = 3t^2 + 2$ then $\frac{dx}{dt} = 3t^2 + 2$.

Thus $$x = \int (3t^2 + 2)\,dt$$

$$= t^3 + 2t + c.$$

We know that initially, i.e. when $t = 0$, $x = 10$.

$\therefore$ $$10 = (0)^3 + 2(0) + c$$

i.e. $$c = 10.$$

Thus $$x = t^3 + 2t + 10.$$

$\therefore$ When $t = 5$, $$x = 5^3 + 2(5) + 10$$

$$= 145.$$

When $t = 5$ the displacement from O is 145 metres.

ISBN 9780170350495

Exercise 8B

Questions **1** to **6** all involve rectilinear motion with x metres and v m/s being the displacement and velocity of a body respectively, relative to an origin O, at time t seconds.

1 If $v = 6t^2 + 4$ find the displacement when $t = 2$ given that for $t = 1, x = 5$.

2 If $v = 10$ find the displacement when $t = 2$ given that for $t = 1, x = 24$.

3 If $v = 10t - 23$ find the times when the body is at the origin given that when $t = 1, x = -6$.

4 If $v = t(t + 2)$ and when $t = 3, x = 13$, find the initial displacement.

5 If $v = 6t - 18$ find the velocity when the body is at the origin, O, given that at the time $t = 3, x = -3$.

6 If $v = 4t - 10$, and when $t = 0$ the body is at the origin, O, find

a when the body is next at O

b the displacement of the body when the velocity is zero.

7 A particle travels along a straight line with its velocity at time t seconds given by v m/s where $v = 4t + 1$. Find the distance travelled in the fifth second.

8 A particle travels along a straight line with its velocity at time t seconds given by v m/s where $v = 9 - 2t$. Find the distance travelled in the fifth second.

Miscellaneous exercise eight

This miscellaneous exercise may include questions involving the work of this chapter, the work of any previous chapters, and the ideas mentioned in the Preliminary work section at the beginning of the book.

1 For each of the following, without using a calculator, write the coordinates of the point where the graph meets the y-axis.

a $y = 2x^4 + x^3 + 3x^2 - 2x - 6$

b $y = 15 + 2x - 7x^2 - 2x^3$

c $y = \dfrac{12}{x + 2}$

d $y = \sqrt{3x + 9}$

e $y = \dfrac{2x + 8}{x - 2}$

f $y = (x - 1)(x + 1)(x - 3)(x + 4)$

2 For each of the following, without using a calculator, write the coordinates of the point(s) where the graph meets the x-axis.

a $y = 2x - 6$

b $y = 6 + 2x$

c $y = x^2 - 9$

d $y = (2 - x)(x - 7)$

e $y = (x - 3)(x + 2)(2x - 7)$

f $y = (x - 1)(x + 1)(x - 3)(x + 4)$

Express each expression of questions **3** to **14** as a power of 2. (I.e. in the form 2^n.)

3 $2 \times 4 \times 8 \times 16 \times 32$ **4** $\frac{1}{16}$ **5** $\frac{1}{2 \times 4 \times 8}$ **6** 0.5

7 0.25 **8** $2^7 \times 2^3$ **9** $2^7 \times 2^3 \times 2 \times 4$ **10** $2^7 \times 2^3 \div 16$

11 8^2 **12** $2 + 2 + 2 + 2$ **13** 1 **14** $(2^6 \div 2^2)^2$

Determine the value of n in each of questions **15** to **23**.

15 $a^4 \div a^9 = a^n$ **16** $a^9 \div a^n = a^4$ **17** $a^n \div a^9 = a^4$

18 $(a^2)^3 = a^n$ **19** $(\sqrt{a})^n = a^5$ **20** $\sqrt{a} \times \sqrt[3]{a} = a^n$

21 $(a^2)^n \times a = a^7$ **22** $\frac{a^9 \div a^n}{a^3 \times a^2} = a^3$ **23** $\frac{\sqrt{a^5}}{\sqrt{a}} = a^n$

State the first five terms for each of the sequences given in numbers **24** to **29**.

24 $T_{n+1} = T_n + 6, T_1 = 15.$ **25** $T_{n+1} = T_n - 7, T_1 = 100.$

26 $T_n = 5T_{n-1}, T_1 = 4.$ **27** $T_{n+1} = 4T_n, T_3 = 96.$

28 $\ldots, T_6 = 243, T_7 = 249, T_8 = 255, T_9 = 261, T_{10} = 267, \ldots$

29 $\ldots, T_7 = 2916, T_8 = 4374, T_9 = 6561, T_{10} = 9841.5, T_{11} = 14762.25, \ldots$

Shutterstock.com/tomertu

30 If the angles of a triangle are in arithmetic progression, show that one of the angles must be 60°.

31 Determine $\frac{dy}{dx}$ for each of the following.

a $y = 3x^2$ **b** $y = 1 + 5x^3$ **c** $y = 0.5x^2 + 3x - 2$

d $y = (1 + 3x)(5x - 2)$ **e** $y = (1 + 3x)^2$ **f** $y = (1 + x)(1 - x)$

ISBN 9780170350495

32 Differentiate each of the following with respect to x.

a $7x^2$ **b** $2x^3 + 5$

c $3x^4 + x^3 - 5x^2 + 9x - 2$ **d** $(3x - 2)(x^2 + 1)$

33 Find the gradient of $y = 2x^3 + x$ at the point $(-1, -3)$.

34 If $f(x) = 3x^2 + 2x + 5$ determine

a $f(1)$ **b** $f(-1)$ **c** $f'(x)$ **d** $f'(2)$

35 Find the point(s) on the following curves where the gradient is as stated.

a $y = 5x^2$. Gradient of 5. **b** $y = 5x^3$. Gradient of 60.

c $y = x^2 + 3x$. Gradient of 7. **d** $y = x^3 - 3x^2$. Gradient of 24.

36 Determine the gradient of

a $y = \frac{12}{x^2}$ at (2, 3), **b** $y = \frac{6}{\sqrt{x}}$ at (4, 3).

37 At what point on the quadratic $y = x^2 - 4x - 1$ is the tangent parallel to the straight line $y = -2x + 5$?

38 Find the equation of the line tangential to the curve $y = x^2$ and perpendicular to the line $2y = x + 6$.

39 An object leaves a point P and travels directly away from P, in a straight line, with its distance from P, in metres, t seconds later, as given by the curve in the graph on the right.

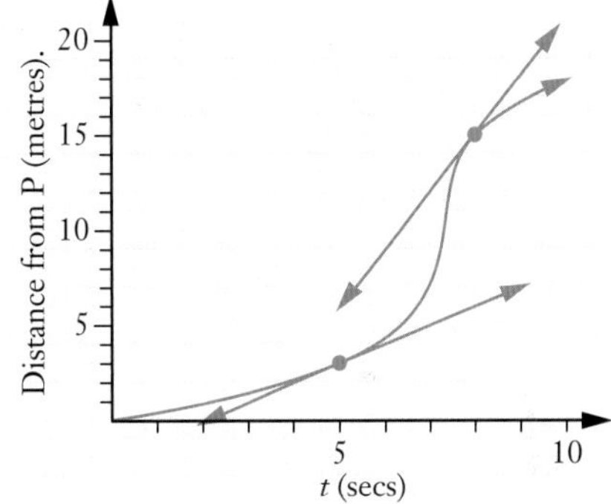

The graph also shows the tangents to the curve at $t = 5$ and $t = 8$. Use the graph to determine

a the average speed of the object for the interval $t = 0$ to $t = 5$,

b the average speed of the object for the interval $t = 5$ to $t = 8$,

c the speed of the object when $t = 5$,

d the speed of the object when $t = 8$.

40 The displacement of a body from an origin O, at time t seconds, is x metres where

$$x = 3t^2 - 12t + 1.$$

a Find the initial velocity of the body.

b Find the initial speed of the body.

c The value of t for which the body has a velocity of 3 m/s.

d The values of t for which the body has a speed of 3 m/s.

41 The graph of $y = ax^2 + bx - 2$ passes through the point $(-3, 10)$ and its gradient at that point is -13. Find a and b and determine the coordinates of the point on the curve where the gradient is -1.

42 Find the equation of the tangent to $y = x - \frac{4}{x}$ at the point (4, 3).

43 Given that $f(x) = ax^3 + bx^2$, $f(2) = -4$ and $f'(3) = 99$, find $f(x)$, $f(3)$ and $f'(2)$.

44 For $f(x) = 2x^3 - 5x + 3$, and using your calculator if you wish, determine

a $f(12), f(17)$ and $f(49)$.

b $f'(12), f'(17)$ and $f'(49)$.

45 Find the following indefinite integrals.

a $\int 60\,dx$

b $\int 60x\,dx$

c $\int 60x^2\,dx$

d $\int 60x^3\,dx$

e $\int 60x^4\,dx$

f $\int 60x^5\,dx$

g $\int (8x^3 - 15x^2 + 2)\,dx$

h $\int (4 - 3x + 2x^2 - x^3)\,dx$

i $\int (x-3)(x+3)\,dx$

j $\int 24x^2(2x-1)\,dx$

46 Find y as a function of x given that

a $\frac{dy}{dx} = 4x - 3$ and when $x = 2, y = 5$.

b $\frac{dy}{dx} = 6x^2 - 2x + 4$ and when $x = -1, y = 0$.

c $\frac{dy}{dx} = 8x^3 - 12x^2 - 4x + 11$ and when $x = 2, y = 4$.

47 Given that $\frac{dy}{dx} = 10x^4 - 6x + 1$, and that when $x = -1, y = 4$, find y when $x = 2$.

48 The radius of a sphere is increasing in such a way that the volume, V cm^3, at time t seconds is given by $V = 7500 + 5400t - 450t^2 + \frac{25t^3}{2}$ for $0 \le t \le 12$.

Calculate

a the volume when $t = 0$,

b the volume when $t = 12$,

c an expression for the instantaneous rate of change in the volume with respect to time,

d the rate at which the volume is increasing (in cm^3/s) when $t = 2$, when $t = 4$ and when $t = 10$.

49 For $y = 3x^2 + 2x + 1$ determine:

a by how much y changes when x changes from $x = 0$ to $x = 10$.

b the average rate of change in y, per unit change in x, when x changes from $x = 0$ to $x = 10$.

c the instantaneous rate of change of y, with respect to x, when $x = 0$.

d the instantaneous rate of change of y, with respect to x, when $x = 10$.

ISBN 9780170350495

50 One of the graphs A to D shown below has $\frac{dy}{dx} = (1 - x)(x + 3)$. Which one?

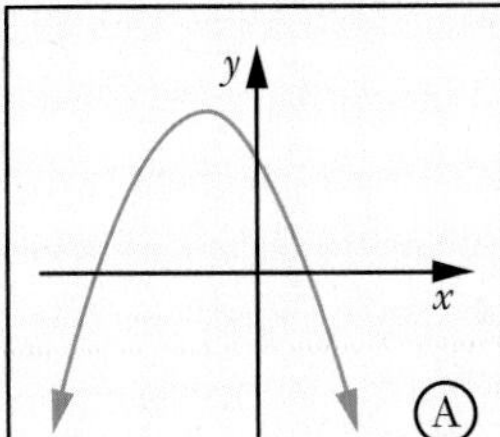

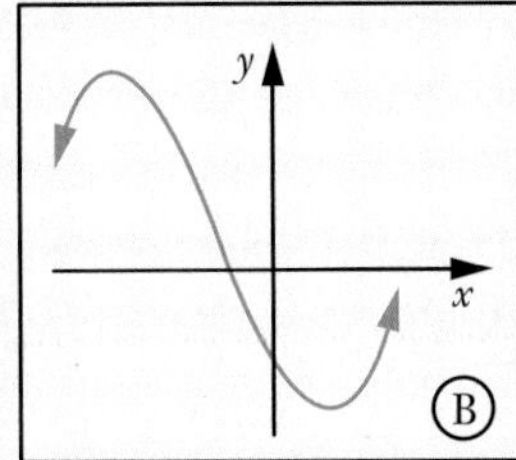

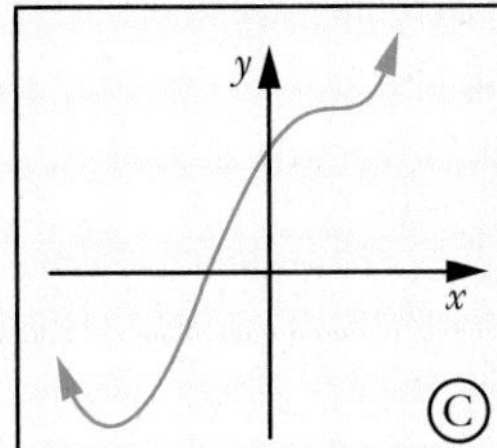

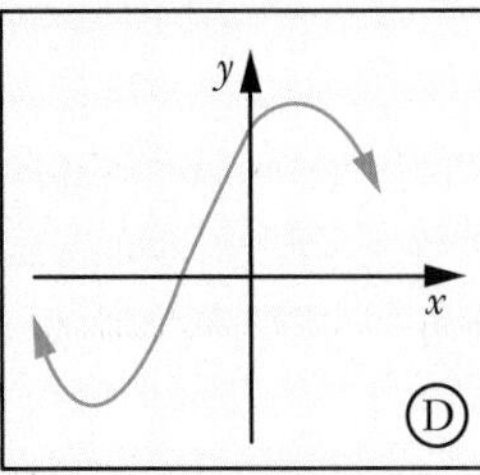

51 A company manufacturing toys wishes to launch a new product but is not sure what price to charge for it. If they charge more they will make more profit on each one they sell but will they sell less because of the higher price?

The minimum they anticipate charging for each of the new toys is \$15.

Market research indicates that if they charge \$(15 + x) the 'demand curve', (likely sales figure, S thousand, plotted against x) has equation:

$$S = 6 + 6x - x^2.$$

Clearly showing your use of calculus, determine the price the company should charge for each toy to maximise likely sales and find what this maximum sales figure would be.

52 An object moves in a straight line such that its distance, x metres, from some point S on the line, at time t seconds, is given by $x = 0.25t^3 + 3$. Find

- **a** the distance the object is from S when $t = 2$,
- **b** the distance the object is from S when $t = 6$,
- **c** the average rate of change in the distance from S from $t = 2$ to $t = 6$,
- **d** the instantaneous rate of change of x per second when $t = 2$,
- **e** the instantaneous rate of change of x per second when $t = 6$.

53 A group of go kart enthusiasts wish to purchase some land for a go kart track (see diagram).

The perimeter of the track is to be 500 metres.

- **a** Find an expression for x in terms of r (and π).
- **b** Find an expression for A, the area enclosed by the track, in terms of r.
- **c** Use differentiation to show that for A to be a maximum the track should be made circular.

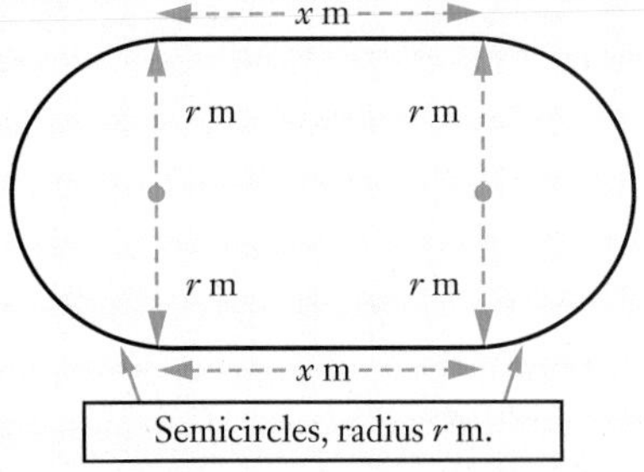

54 On 1 January 2015 Nadine opens an account by depositing \$5000 in an account earning interest at 6.8% compounded annually. On every 1 January thereafter Nadine adds a further \$800 to the account. If the interest remains at 6.8% throughout how much is the account worth when Nadine closes it on 31 December 2025?

ISBN 9780170350495

55 The organisers of a scheme to raise funds to send a group of dancers to an interstate competition plan to sell boxes of cakes. They estimate that if they charge \$3 per box they will sell 120 boxes. For each five cents that they raise the price per box they expect their sales to decrease by one box. Suppose they raise the price by 'x' lots of five cents i.e. by $5x$ cents.

The price per box will then be $\$(3 + 0.05x)$.

a Write down an expression, in terms of x, for the number of boxes they would then expect to sell.

b Write down a revenue function, $R(x)$, for the total revenue (i.e. total income before costs are deducted) for selling this number of boxes at this price.

c What price should they sell each box for to maximise the revenue?

d What would this maximum revenue be?

It costs the organisers \$2 to produce each box of cakes.

e Write down a profit function, $P(x)$, for the total profit they could expect for selling the cakes at $\$(3 + 0.05x)$ per box.

f For what price should they sell each box to maximise their profit?

g What would this maximum profit be?

56 The displacement of a body from an origin O, at time t seconds, is x metres where x is given by:

$$x = 10t - t^2 + 4.$$

Find **a** the displacement when $t = 7$

b the value of t when the body is at rest

c the distance the body travels from $t = 1$ to $t = 7$.

57 The final sections of a big dipper ride at an amusement park has the shape shown on the right. The sections, A → B, B → C and C → D, are quadratic functions. At B and at C, the points where one quadratic function flows into the next, there are no 'gaps' between the functions and the gradient at the end of one function is the same as the gradient at the beginning of the next.

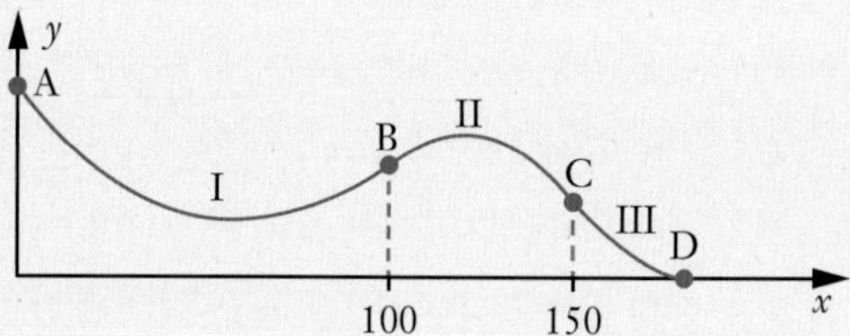

- For section I, i.e. A → B the equation is $y = 0.01x^2 - 1.2x + 50$.
- For section II, i.e. B → C the equation is $y = ax^2 + bx - 250$.
- For section III, i.e. C → D the equation is $y = cx^2 + dx + 605$.
- Point A has coordinates $(0, e)$.
- Point B has coordinates $(100, f)$.
- Point C has coordinates $(150, g)$.
- Point D has coordinates $(h, 0)$.

Determine a, b, c, d, e, f, g and h.

ISBN 9780170350495

ANSWERS

Exercise 1A PAGE 8

1 36 **2** 8 **3** 16
4 1 **5** 1 **6** 2
7 1 **8** 2 **9** 1
10 8 **11** 1 **12** –1
13 –32 **14** 1 **15** –1
16 4 **17** 5 **18** 2
19 3 **20** 9 **21** $\frac{1}{9}$
22 $\frac{1}{4}$ **23** $\frac{1}{8}$ **24** $\frac{3}{4}$
25 $\frac{1}{6}$ **26** 7 **27** 5
28 $\frac{1}{5}$ **29** 1 **30** 1
31 1 **32** –2 **33** 125
34 $\frac{4}{3}$ **35** $\frac{2}{3}$ **36** 27
37 $\frac{1}{27}$ **38** $\frac{1}{16}$ **39** $\frac{3}{2}$
40 25 **41** 25 **42** $\frac{1}{16}$
43 $\frac{1}{512}$ **44** 2^{16} **45** 2^{13}
46 2^6 **47** 2^{14} **48** 2^8
49 2^0 **50** 2^{-1} **51** 2^{-3}
52 2^{-3} **53** 3^3 **54** 3^4
55 3^0 **56** 3^{-1} **57** $3^{0.5}$
58 $3^{0.25}$ **59** 3^{-3} **60** $3^{-0.5}$
61 3^6 **62** 10^2 **63** 10^{-1}
64 10^{-1} **65** 10^{-2} **66** 10^{-2}
67 10^0 **68** 10^6 **69** 10^6
70 10^6 **71** 10^9 **72** 10^{-3}
73 $10^{0.5}$ **74** 10^3 **75** $10^{-0.5}$
76 $10^{\frac{3}{2}}$ **77** 5 **78** 4
79 3 **80** 4 **81** 7
82 5 **83** 8 **84** 8
85 10 **86** 8 **87** 9
88 8 **89** 10 **90** 8
91 7 **92** 3 **93** 2
94 2 **95** 15 **96** 3
97 4 **98** a^7 **99** $\frac{1}{a^3}$
100 $\frac{1}{b^5}$ **101** b^6 **102** $\frac{1}{b^6}$
103 a^2 **104** $\frac{1}{a^6}$ **105** a^2
106 b^{10} **107** b^{18} **108** $-4a^5$
109 $16a^5$ **110** $\frac{1}{a^2}$ **111** $\frac{a^9}{b^3}$
112 $6a^3$ **113** $\frac{1}{a^3}$ **114** a^2
115 $2a$ **116** $8a^5$ **117** $32a^5$
118 $4a^2b^4$ **119** $\frac{10y^2}{a^5}$ **120** $\frac{3}{2a^4b^3}$
121 $\frac{y}{3x}$ **122** $-9a^8b$ **123** a^2b^5
124 $\frac{3a^4}{2b^2}$ **125** $\frac{1}{a^5}$ **126** $\frac{y^3}{x^4}$
127 $\frac{b^4}{a^8}$ **128** $a^4 + a$ **129** $a^2 + a^4$
130 $2a + 3$ **131** $\frac{4}{5}$ **132** $7(2^n)$
133 9

Exercise 1B PAGE 13

1 $x = 3$ **2** $x = 5$ **3** $x = 7$
4 $x = -3$ **5** $x = -5$ **6** $x = -7$
7 $a = 0.5$ **8** $a = -0.5$ **9** $y = -2$
10 $c = 3$ **11** $d = 3$ **12** $x = 3$
13 $x = 3$ **14** $x = 6$ **15** $x = 2$

ISBN 9780170390408

16 $x = 1$ **17** $x = 0.75$ **18** $x = -1$
19 $x = 3$ **20** $x = 2$ **21** $x = 5$
22 $k = 0.75$ **23** $p = -0.25$ **24** $x = 1.5$
25 $x = 1$ **26** $h = 6$ **27** $x = 0.5$
28 $x = 1$ **29** $x = -0.25$ **30** $n = 2$
31 $a = \pm 4$ **32** $p = \pm 10$ **33** $x = 2$
34 $x = 4$ **35** $x = 16$ **36** $x = 64$
37 $h = 0.25$ **38** $y = 0.5$ **39** $p = 3$
40 $x = 10\,000$ **41** $x = \pm 5$ **42** $x = \pm\frac{2}{3}$
43 $x = \pm 3$ **44** $x = 1000$ **45** $p = \pm 4$
46 $x = 4$ **47** $x = 81$ **48** $x = \frac{1}{16}$
49 $w = 4$ **50** $x = 2$ **51** $h = 24$
52 $x = -1$ or 7 **53** $w = \frac{9}{4}$, i.e., 2.25
54 $z = \frac{27}{8}$, i.e., 3.375
55 **a** The third solution is $x = 0$.
b $x = 0, x = -4, x = 4$
c **i** $x = 0, x = 4$ **ii** $x = 0, x = -5, x = 5$
iii $x = 0, x = 25$ **iv** $x = 0$

Exercise 1C PAGE 16

1 $x = 4.5$ **2** $x = 3.2$
3 $x = 2.4$ **4** $x = 2.8$
5 $x = 6.7$ **6** $x = 2.2$
7 $x = 2.46$ **8** $x = 2.55$
9 $x = 3.32$ **10** $x = 6.26$
11 $x = 3.86$ **12** $x = 3.53$
13 $x = 8.43$ **14** $x = -1.51$
15 $x = 4.35$ **16** $x = 3.34$
17 $x = 2.74$ **18** $x = -6.99, x = 3.38$

Miscellaneous exercise one PAGE 17

1 Completed table not given here. $y = -x^3 + 2x^2 - x + 7$
2 **a** Australia has an area of approximately 7 682 000 km^2.
b Light travels at a speed of 300 000 000 m/sec.
c A golf ball has a mass of approximately 0.045 kg.
d The Earth is approximately 150 000 000 km from the sun.
e Gamma waves have a wave length less than 0.000 000 000 01 metres.
f The Earth orbits the sun at a speed of approximately 107 000 km/hr.
g In 1961 the first man in space, Yuri Gagarin, flew his spacecraft at a speed of 27 400 km/hr, i.e. approximately 7600 m/sec.
3 **a** At the beginning of this century China had a population of approximately 1.27×10^9 and India had a population of approximately 1.03×10^9.
b The egg cell, or ovum, with a radius of approximately 5×10^{-5} metres, i.e. 5×10^{-2} mm, is the largest single human cell.
c It is thought that approximately 1.1×10^6 people die each year of malaria.
d Some adult wasps of a particular species could weigh just 5×10^{-3} grams.
e Concorde, the first supersonic passenger airliner, had a cruising speed of 2.16×10^3 km/hr.
4 **a** 12 000 000 **b** 46 800
c 305 000 000 **d** 0.01
e 0.206 **f** 0.006
5 **a** 5^2 **b** 5^3 **c** $5^{0.5}$
d 5^{-1} **e** 5^{-2} **f** $5^{-0.5}$
g 5^7 **h** 5^7 **i** 5^6
j 5^{12} **k** 5^6 **l** 5^0
m 5^5 **n** 5^2 **o** 5^5
6 **a** $x = 5$ **b** $x = 4$
c $x = -2$ **d** $x = \frac{5}{2}$
e $x = \frac{5}{3}$ **f** $x = \frac{1}{3}$
g $x = -\frac{1}{3}$ **h** $x = -3$
i $x = -\frac{2}{5}$ **j** $x = \frac{1}{4}$
k $x = -\frac{3}{2}$ **l** $x = 8$
7 **a** $y = \pm\frac{1}{3}$ **b** $p = \frac{4}{9}$
c $x = \frac{3}{2}$ **d** $x = 8$
e $x = \pm 5$ **f** $t = \pm\frac{1}{5}$
g $t = \frac{27}{2}$ **h** $x = 9$
i $x = 0, x = 1$ **j** $x = 0, x = \pm\pm 1$
8 **a** $x = 1, x = 3$ **b** $x = 0, x = 2$
c $x = -1, x = 0$ **d** $x = 1, x = 3$

ISBN 9780170390408

Exercise 2A PAGE 25

1 Rule: $y = 3^x$

x	0	1	2	3	4	5
y	1	3	9	27	81	243

2 Rule: $y = 7^x$

x	0	1	2	3	4	5
y	1	7	49	343	2401	16 807

3 Rule: $y = 1.5 \times 2^x$

x	0	1	2	3	4	5
y	1.5	3	6	12	24	48

4 Rule: $y = 1.75 \times 8^x$

x	0	1	2	3	4	5
y	1.75	14	112	896	7168	57 344

5 Rule: $y = 2^{x+1}$

x	0	1	2	3	4	5
y	2	4	8	16	32	64

6 Rule: $y = 2.5 \times 4^{x+1}$

x	1	2	3	4	5	6
y	40	160	640	2560	10 240	40 960

7 **a** Quadratic **b** $y = x^2 + 1$
8 **a** Exponential **b** $y = 4^x$
9 **a** Linear **b** $y = 2x + 3$
10 **a** Quadratic **b** $y = 2x^2$
11 **a** Exponential **b** $y = 1.5(8)^x$
12 **a** Exponential **b** $y = 5^x$
13 **a** Quadratic **b** $y = x^2 + x$
14 **a** Exponential **b** $y = 6^x$
15 **a** Exponential **b** $y = 3(2)^x$
16 **a** Reciprocal **b** $y = \frac{60}{x}$
17 **a** Cubic **b** $y = x^3 + 1$
18 **a** Linear **b** $y = 20 - 3x$
19 **a** (0, 1)
b Discuss your answer with your teacher.
20–22 Discuss your answer with your teacher.
23 **a** $y = 2^x$ **b** $y = 3^x$
24 **a** 12 **b** 28 **c** 40
25 **a** $y = 3^{x-2}$ **b** $y = 2^x + 2$
c $y = 2^{x-2}$ **d** $y = 3^x - 2$
e $y = 3^{x+1} + 2$ **f** $y = 2^{x-2} - 2$

Exercise 2B PAGE 31

3 Approximately 61 million, assuming annual growth rate for the given years continues.
4 Approximately 6000.
5 **a** $A = 80, k = 1.08$
b 8%
c 1200
6 Approximately 1200.
7 **a** 68 **b** 29 **c** 0.84, 115
d 115 **e** 14
8 **a** $k \approx 80, a \approx 0.92$
b Approximately 27.
c Approximately 2021.
9 **a** 10 000 of A and 1000 of B.
b 4200 of A and 1300 of B.
c 6.2
10 **a** $k \approx 850$, a is between 0.9 and 0.91.
b Approximately 14 weeks.

Miscellaneous exercise two PAGE 35

1 **a** II **b** IV **c** III
d I **e** III **f** IV
g III **h** I
2 **a** $x = \pm 7$ **b** $x = \pm 10$ **c** $x = 10$
d $x = 2$ **e** $x = 4$ **f** $x = 0$
g $x = 3$ **h** $x = -1$ **i** $x = -2$
j $x = -3$ **k** $x = -1$ **l** $x = -2$
m $x = -3$ **n** $x = 0, \pm 5$ **o** $x = -0.125$
p $x = 0.5$ **q** $x = 0.25$ **r** $x = \pm 3$
3 **a** 12 000 **b** 12 610 000 **c** 0.000 26
d 6 **e** 12 630
4 **a** $y = 2^{x+3}, y = 8 \times 2^x$ **b** $y = 3^x - 2$
5 Check reasonableness of answers by evaluating $5^{1.6}$, $5^{2.4}$ and $5^{2.5}$ on a calculator.
6 –2 or 0
7 0 or 2
8 **a** $k \approx 18.9, a \approx 0.93$
b Approx. 6.22 a.m. (Remember, graph shows number of *half* hours.)

ISBN 9780170390408

Exercise 3A PAGE 42

1 18	**2** 26	**3** 44
4 38	**5** 42	**6** 36
7 72	**8** 44	**9** 42
10 46	**11** 324	**12** 2744
13 8	**14** 20	**15** 28
16 26	**17** 29	**18** 21
19 27	**20** 9	**21** 162
22 18	**23** 26	**24** 4374
25 27	**26** 216	**27** 343
28 91	**29** 4	**30** 7
31 49	**32** 94	

Exercise 3B PAGE 48

1 $T_1 = 6, T_{n+1} = T_n + 4$.

2 $T_1 = 28, T_{n+1} = T_n - 2$.

3 $T_1 = 5, T_{n+1} = T_n + 10$.

4 $T_1 = 7.5, T_{n+1} = T_n + 2.5$.

5 $T_1 = 100, T_{n+1} = T_n - 11$.

6 $T_1 = 6, T_n = 2T_{n-1}$.

7 $T_1 = 0.375, T_n = 4T_{n-1}$.

8 $T_1 = 384, T_n = 0.25T_{n-1}$.

9 $T_1 = 50, T_n = 3T_{n-1}$.

10 $T_1 = 1000, T_n = 1.1T_{n-1}$.

11 Neither arithmetic nor geometric.

12 Geometric.

13 Arithmetic.

14 Arithmetic.

15 Neither arithmetic nor geometric.

16 Geometric.

17 Geometric.

18 Arithmetic.

19 Neither arithmetic nor geometric.

20 Neither arithmetic nor geometric.

21 Arithmetic.

22 Geometric.

23 $T_1 = 8, T_2 = 11, T_3 = 14, T_4 = 17, T_{n+1} = T_n + 3$.

24 $T_1 = 100, T_2 = 97, T_3 = 94, T_4 = 91, T_{n+1} = T_n - 3$.

25 $T_1 = 11, T_2 = 22, T_3 = 44, T_4 = 88, T_{n+1} = 2T_n$.

26 $T_1 = 2048, T_2 = 1024, T_3 = 512, T_4 = 256, T_{n+1} = 0.5T_n$.

27 **b** $N_{n+1} = N_n + 800$

28 **a** The sequence is a geometric progression.

b The next three terms after the first are 550, 605 and 665.5.

29 **a** The sequence is a geometric progression.

b The next three terms after the first are 1250, 1562.5 and 1953.125.

30 **a** The sequence is a geometric progression.

b The next three terms after the first are 21 600, 19 440 and 17 496.

31 **a** $T_1 = 3, T_{n+1} = T_n + 5$.

b The sequence is arithmetic.

32 **a** $T_1 = 1.5, T_{n+1} = T_n \times 2$.

b The sequence is geometric.

33 **a** $T_1 = 4, T_2 = 9, T_3 = 16, T_4 = 25, T_5 = 36$.

b Neither arithmetic nor geometric.

34 **a** After 1 year, 2 years, 3 years and 4 years the account is worth \$1296, \$1392, \$1488 and \$1584 respectively.

b The amounts are in arithmetic progression.

c $T_1 = \$1200, T_{n+1} = T_n + \$96$.

35 $T_1 = 4, T_{n+1} = T_n + 1$.

36 $T_1 = \$45\,000, T_{n+1} = T_n + \$1500$. The terms of the sequence progress arithmetically.

37 \$68 000 in 2014, \$71 400 in 2015, \$74 970 in 2016, \$78 718.50 in 2017.

$T_1 = \$68\,000, T_{n+1} = 1.05T_n$.

38 $T_1 = \$1500, T_{n+1} = 1.08T_n$.

39 $T_1 = \$36\,000, T_{n+1} = 0.85T_n$.

Exercise 3C PAGE 58

1 $T_{100} = 506$	**2** $T_{100} = 289$
3 $T_{100} = 815$	**4** $T_{100} = -120$
5 $T_{25} = 5 \times 2^{24}$	**6** $T_{25} = 1.5 \times 4^{24}$
7 $T_{25} = 8 \times 3^{24}$	**8** $T_{25} = 11 \times 2^{24}$
9 $T_{28} = 223$	**10** $T_{20} = 3\,495\,265$
11 $T_{19} = 774\,840\,977$	**12** $T_{45} = 6$

13 $T_1 = 48, T_{n+1} = T_n + 3$. Julie successfully completes 90 items on the 15th day.

14 Substituting y for T_n and x for n the rule $T_n = a + (n - 1)d$ becomes $y = dx + (a - d)$.

This is the equation of a straight line with gradient d, cutting the y-axis at $(0, a - d)$.

15 Substituting y for T_n and x for n the rule

$T_n = ar^{n-1}$ becomes $y = ar^{x-1}$, i.e. $y = \left(\frac{a}{r}\right) r^x$.

An exponential function cutting the y-axis at $\left(0, \frac{a}{r}\right)$.

ISBN 9780170390408

16 As $n \to \infty$, the 'nd' term in the expression $a + (n - 1)d$ will dominate.

Thus as $n \to \infty$, T_n will be increasingly large and positive if $d > 0$ and increasingly large and negative if $d < 0$.

17 As $n \to \infty$, the n in the expression ar^{n-1} will dominate.

Thus as $n \to \infty$,

if $r > 1$, T_n will become increasingly large, either positively or negatively dependent on the sign of the constant a.

if $r < -1$, T_n will become increasingly large, alternating between large negative and large positive.

if $-1 < r < 1$, T_n will become smaller and smaller, maintaining the sign of the constant a if r is positive and alternating between small positive and small negative if r is negative.

That is, if $-1 < r < 1$, as $n \to \infty$, $T_n \to 0$.

18 The first four terms are 8, 11, 14, 17. The 50th term is 155. The 100th term is 305.

19 The first four terms are 100, 97, 94, 91. The 50th term is –47. The 100th term is –197.

20 11, 22, 44, 88, 180 224, 184 549 376

21 2048, 1024, 512, 256, 0.0625

22 **a** $T_n = 9 + (n - 1) \times 6$, i.e. $T_n = 6n + 3$
b $T_n = 7 + (n - 1) \times 1.5$, i.e. $T_n = 1.5n + 5.5$

23 **a** $3 \times 2^{n-1}$ **b** $100 \times 1.1^{n-1}$

24 **a** 856 **b** 3495
c The 142 858th term.

25 **a** 126 953.125 **b** The 14th term.

26 **a** 844 700 **b** The 60th term.

27 1, 8, 27, 64, neither

28 **a** 64 **b** 7

29 **a** 1850 **b** 2000

30 **a** 7 971 615 **b** 5

31 **a** –1 835 008 **b** 7

32 The amount in the account at the end of ten years is \$8635.70.

33 Just after the end of the 22nd year, i.e. early in the 23rd year.

34 $T_{n+1} = 1.08 \times T_n + \$200$, $T_{10} = \$11\,533.01$

35 $T_{n+1} = 1.08 \times T_n - \$200$, $T_{10} = \$5738.39$

Miscellaneous exercise three PAGE 61

1 **a** Quadratic **b** Exponential
c Linear **d** Quadratic
e Reciprocal **f** Linear
g Linear **h** Quadratic
i Quadratic **j** Reciprocal
k Linear **l** Exponential

2 **a** $x \approx 2.3$ **b** $x \approx 2.6$ **c** $x \approx 1.4$

3 **a** 3 **b** –3 **c** –1
d 0.5 **e** 0 **f** 1.5
g –6 **h** 1.5

4 **a** $T_n = 4 \times 1.5^{n-1}$ **b** $T_n = \frac{8}{3} \times 1.5^n$

5 243

6 **a** $x = 3$ **b** $x = -3$ **c** $x = -1$
d $x = 8$ **e** $x = \pm 8$ **f** $x = 11$

7 **a** 4 **b** 64 **c** 9
d 0.2 **e** 2

8 **a** –4, **b** –4, 4, 4 **c** Neither

9 **a** –2, **b** –4, 8, 16 **c** Neither

10 **a** $48 - 4k$ **b** 95.5

11 a^7 **12** $12x^3y^4$ **13** $\frac{3a^2}{2b^2}$

14 $72a^8b^3$ **15** $\frac{9}{8a^4b^3}$ **16** $\frac{48b}{a}$

17 $\frac{2a^5}{b^5}$ **18** $k^4 + 1$ **19** $p^3 - p^6$

20 125 **21** 25 **22** $\frac{8}{3}$

Exercise 4A PAGE 69

1 **a** 68 **b** 100 **c** 138

2 **a** 53 **b** 123 **c** 28

3 **a** –9 **b** 0 **c** 9

4 $T_1 = 6$, $T_2 = 11$, $T_3 = 16$, $T_4 = 21$.
$S_1 = 6$, $S_2 = 17$, $S_3 = 33$, $S_4 = 54$.

5 $T_1 = 11$, $T_2 = 14$, $T_3 = 17$, $T_4 = 20$.
$S_1 = 11$, $S_2 = 25$, $S_3 = 42$, $S_4 = 62$.

6 $T_1 = 22$, $T_2 = 19$, $T_3 = 16$, $T_4 = 13$.
$S_1 = 22$, $S_2 = 41$, $S_3 = 57$, $S_4 = 70$.

7 $T_1 = 25$, $T_2 = 32$, $T_3 = 39$, $T_4 = 46$. $T_5 = 53$. Yes

8 $T_1 = 1$, $T_2 = 4$, $T_3 = 9$, $T_4 = 16$. $T_5 = 25$. No

9 **a** 48 **b** 8780

10 **a** 174 **b** 60

11 5050

12 **b** 2088

13 **b** 14 309

14 78 km, 1470 km **15** 285

16 \$31 500 **17** A: \$762 500 B: \$734 000

18 \$6840

ISBN 9780170390408

Exercise 4B PAGE 73

1 $T_1 = 6,\quad T_2 = 18,\quad T_3 = 54,\quad T_4 = 162.$
$S_1 = 6,\quad S_2 = 24,\quad S_3 = 78,\quad S_4 = 240.$

2 $T_1 = 16,\quad T_2 = 24,\quad T_3 = 36,\quad T_4 = 54.$
$S_1 = 16,\quad S_2 = 40,\quad S_3 = 76,\quad S_4 = 130.$

3 $T_1 = 1,\quad T_2 = 1,\quad T_3 = 2,\quad T_4 = 3,\quad T_5 = 5.$ No

4 $T_1 - 8,\quad T_2 = 24,\quad T_3 = 72,\quad T_4 = 216, T_5 = 648.$ Yes

5 32767 **6** 40940

7 650871 **8** 104139.36

9 3071.25, 12287.25, 49151.25

10 20 **11** 25

12 393216, 524287.5 **13** $1015000

14 a Approximately 5500 tonnes.
b Approximately 6050 tonnes.
c Approximately 6655 tonnes.
d Approximately 107000 tonnes.

15 a Approximately $69000.
b Approximately $79350.
c Approximately $211000.
d Approximately $1218000.

16 Entries in last line of table are:
1/1/18 $\$1200 \times 1.1^4$ $\$1200 \times 1.1^3$ $\$1200 \times 1.1^2$
$\$1200 \times 1.1$ $1200 $7326.12
Immediately following the deposit of $1200 on 1/1/29 there will be $43 140 in the account, to the nearest dollar.

17 $14784

18 a Approximately 2653.
b Approximately 3299.
c Approximately 3299.
d Approximately 43200.
e Approximately 125700.

19 a 17 years
b Approximately 79500 tonnes.

20 a First term $\$P$, common ratio 1.095, number of terms 21.
b 829.7

Exercise 4C PAGE 79

1 GP A: **a** 0.4 **b** S_∞ exists and equals 40.
GP B: **a** 1.5 **b** S_∞ does not exist.
GP C: **a** 0.3 **b** S_∞ exists and equals 50.

2 a S_∞ exists and equals 200.
b S_∞ exists and equals 400.
c S_∞ does not exist.
d S_∞ exists and equals 450.
e S_∞ does not exist.
f S_∞ exists and equals 50.
g S_∞ exists and equals 0.9.
h S_∞ exists and equals 2048.

3 0.6

4 66

5 Table not shown here.
a 25 mg **b** 10 mg

6 250.
The idea that the athlete's performance might diminish according to some geometrical pattern is not unreasonable as he would tire as time went on. Hence the situation could feasibly be modelled by a geometrical sequence but we would be surprised if the numbers exactly fitted the model.
However, if the geometrical sequence were continued, by the 10th minute the athlete is completing the exercise approximately 7 times and by the 15th minute approximately 2 times so it could be argued that there is resting going on in these later minutes. If the athlete has to complete the exercise at least once each minute then counting would stop after about 18 minutes with a total of about 245 completions.

7 a 1.2 m **b** Approx. 9 cm. **c** 8 m

8 11.67 m

Miscellaneous exercise four PAGE 81

1 a 2^6 **b** 2^8 **c** 2^7
d 2^2 **e** 2^{10} **f** 2^2
g 2^{14} **h** 2^0 **i** 2^5

2 $\frac{1}{2}$ **3** 16 **4** $\frac{9}{4}$ (i.e., 2.25)

5 1 **6** 64 **7** $\frac{1}{25}$ (i.e., 0.04)

8 $\frac{1}{3}$ **9** 5 **10** $\frac{1}{7}$

11 Compare your reasoning with that of others in your class.

ISBN 9780170390408

12

	$T_1, T_2, T_3, T_4, T_5, \dots$	Recursively defined
a	17, 22, 27, 32, 37, ...	$T_n = T_{n-1} + 5, T_1 = 17$
b	100, 93, 86, 79, 72, ...	$T_n + _1 = T_n - 7, T_1 = 100$
c	5, 15, 45, 135, 405, ...	$T_n = 3T_{n-1}, T_1 = 5$
d	6, 10, 14, 18, 22, ...	$T_n + _1 = T_n + 4, T_1 = 6$
e	2, 6, 18, 54, 162, ...	$T_n + _1 = 3T_n, T_1 = 2$
f	17, 9, 1, –7, –15, ...	$T_n + _1 = T_n - 8, T_1 = 17$

13 $a = 3, k = 2$, for **a** $T_{20} = 524288$, for **b** $T_{20} = 62$

14 **a** 15

15 **a** $x = 29, \quad T_{n+1} = T_n + 21, \quad T_1 = 8$

b $x = 20, \quad T_{n+1} = 2.5 \times T_n, \quad T_1 = 8$. Or:

$x = -20, \quad T_{n+1} = -2.5 \times T_n, \quad T_1 = 8$.

16 **a** When $t = 3.493$ (to three decimal places), i.e. in approximately 3.5 years.

b When $t = 6.986$ (to three decimal places), i.e. in approximately 7 years.

17 $T_1 = 30, T_{n+1} = T_n + 3$.

One day prior to the championships Rosalyn will practise for 90 minutes.

During the 20 days prior to the championships Rosalyn will practise for a total of 20 hours and 30 minutes.

18 After 20 years, account A will have a balance of \$3 207 135 (nearest dollar) compared to account B, which after 20 years will have a balance of \$1 949 636 (nearest dollar).

The organisers need to have \$607 906 available 'now'. (Rounded up to next dollar.)

Exercise 5A PAGE 88

1 **a** A → B, D → F

b B → D, F → H, H → I

c B, D, F, H

2 I: B, C, D, E, F, G — II: A

III: H — IV: A, D, G

V: A, D, H — VI: C, F

VII: A, E, G — VIII: A, B, F, G

IX: A, E, F, G — X: B, F, H

3 **a** C, E, H, K, M, O. **b** A, B, I, J, N, P.

c D, F, G, L.

4 **a** 2 **b** 4 **c** 0 **d** –2

e –4 **h** 2 **g** 2

5

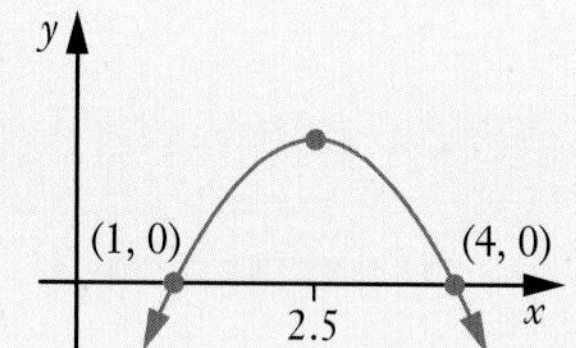

6

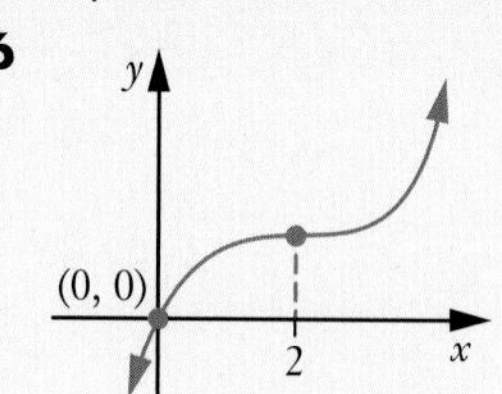

7

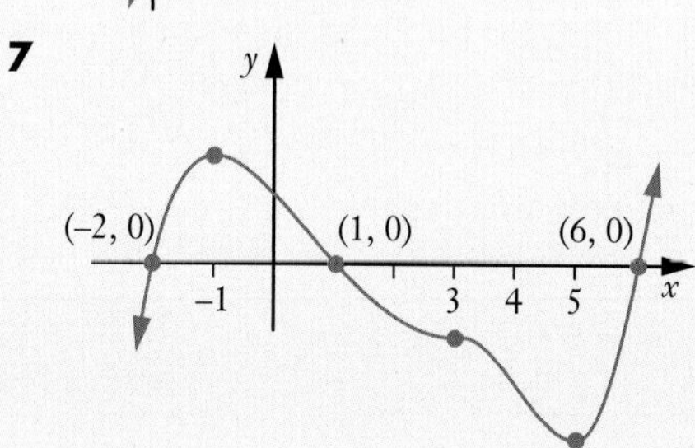

Exercise 5B PAGE 92

1

Point P	Point Q	Grad of chord PQ
(2, 4)	(4, 16)	6
(2, 4)	(3, 9)	5
(2, 4)	(2.5, 6.25)	4.5
(2, 4)	(2.1, 4.41)	4.1
(2, 4)	(2.01, 4.0401)	4.01
(2, 4)	(2.001, 4.004001)	4.001
(2, 4)	(2.0001, 4.00040001)	4.0001

Thus the gradient of $y = x^2$ at $x = 2$ is 4.

2 $y - x^2$

x	0	1	2	3	4	5
grad	0	2	4	6	8	10

Compare your rule with those of others.

3 $y = 3x^2$

x	0	1	2	3	4	5
grad	0	6	12	18	24	30

Compare your rule with those of others.

ISBN 9780170390408

Exercise 5D PAGE 101

1 $2x$ 2 $3x^2$ 3 1
4 $4x^3$ 5 0 6 $12x$
7 $24x^3$ 8 7 9 16
10 $14x^6$ 11 $14x$ 12 9
13 $\frac{x}{5}$ 14 $4x^5$ 15 $9x^5$
16 $2x^6$ 17 $8x$ 18 $20x^3$
19 $24x^2$ 20 0 21 $7x^6$
22 $24x^5$ 23 $18x$ 24 5
25 0 26 $18x^2$ 27 $32x^3$
28 $15x^4$ 29 $6x^5$ 30 $42x^6$
31 $16x^3$ 32 10 33 12
34 12 35 12 36 80
37 7 38 −20 39 2
40 0.8 41 (1, 1) 42 (1, 1), (−1, −1)
43 (1.5, 6.75) 44 (0.5, 0.25), (−0.5, −0.25)
45 (1, 1) 46 (−1, 1)
47 $y = 6x - 4$ 48 $y = -6x - 3$
49 $y = 20x - 20$ 50 $y = -20x - 20$
51 $y = 16x - 24$ 52 $y = 18x - 72$
53 **a** 24 **b** −3 **c** $9x^2$ **d** 36
54 **a** 6 **b** 24 **c** $3x$ **d** 6
55 **a** y changes by 234 units (from 16 to 250) when x changes from $x = 2$ to $x = 5$.
b y changes at an average rate of 78 units per unit change in x when x changes from 2 to 5.
c When $x = 2$ the instantaneous change in y is 24 units per unit change in x.
d When $x = 5$ the instantaneous change in y is 150 units per unit change in x.
56 (−1, 8) gradient −16, (2, 32) gradient 32.
57 (−2, −8) gradient 12, (0, 0) gradient 0, (2, 8) gradient 12.
58 $\frac{1}{54}$, 1.5
59 $-\frac{1}{6}, \frac{1}{6}$.

Exercise 5E PAGE 104

1 $2x + 3$ 2 $3x^2 - 4$
3 $12x - 21x^2$ 4 $12x^3 + 6x^2 - 5$
5 $7 + 2x$ 6 $12x - 3$
7 $8x + 7$ 8 $15x^2 - 8x$
9 $20x^3 - 3$ 10 $4x + 7$
11 $-6x + 7$ 12 $1 + 2x + 3x^2 + 4x^3$
13 $-4 + 6x - 6x^2 + 4x^3$ 14 −3
15 24 16 21
17 11 18 $y = 7x - 4$
19 $y = 13x - 50$ 20 $y = 8x - 34$
21 $y = 1$ 22 (−5, 76), (1, −2)
23 (−3, 0) gradient −8, (5, 0) gradient 8.
24 (5, −10) 25 (−3, 20), (1, −4)

Exercise 5F PAGE 107

1 $\frac{1}{2\sqrt{x}}$ 2 $-\frac{1}{x^2}$ 3 $-\frac{3}{x^2}$
4 $\frac{3}{\sqrt{x}}$ 5 $\frac{2}{x^{\frac{2}{3}}}$ 6 $\frac{3}{2}\sqrt{x}$
7 $\frac{2}{3x^{\frac{2}{3}}}$ 8 $-\frac{3}{x^4}$ 9 $-\frac{4}{x^5}$
10 $-\frac{6}{x^4}$ 11 $-\frac{20}{x^5}$ 12 $2x + \frac{1}{2\sqrt{x}}$
13 $6x - \frac{2}{\sqrt{x}}$ 14 $1 - \frac{1}{x^2}$ 15 $2x + \frac{2}{x^3}$
16 $\frac{1}{2\sqrt{x}} - \frac{3}{x^2}$ 17 $2x + 1 - \frac{1}{x^2} - \frac{2}{x^3}$
18 $-\frac{2}{x^2}$ 19 $-\frac{3}{2\sqrt{x^3}}$ 20 $-\frac{2}{x^{\frac{4}{3}}}$
21 $-\frac{1}{3x^{\frac{4}{3}}}$ 22 −5 23 0.25
24 0.0625 25 11 26 $\frac{8}{3}$
27 −4 28 −0.5
29 (2, 0.5), (−2, −0.5) 30 (0.25, 0.5)
31 (9, −243) 32 $y = 0.25x + 1$
33 $y = -x + 2$ 34 $y = -0.25x + 0.75$
35 $(-1\frac{1}{3}, -1\frac{11}{12}), (1\frac{1}{3}, 1\frac{11}{12})$
36 Answers not given here. Discuss with others in your class and your teacher.

Miscellaneous exercise five PAGE 109

1 **a** 4 **b** 16 **c** 3 **d** 7
e 9 **f** 4 **g** 4 **h** 7
i 3 **j** 3 **k** 0 **l** 4
m 2 **n** 4 **o** 9

ISBN 9780170390408

2 **a** 9 **b** 16

3 **a** 13 **b** −24

4 **a** 1024 **b** 1 073 741 824

c 2046 **d** 2 147 483 646

5 **a** $-3x^2$ **b** $10x - \frac{3}{\sqrt{x}}$ **c** $10x - \frac{2}{x^3}$

6 −4

7 **a** Reciprocal. $y = \frac{-6}{x}$

b Quadratic. $y = x^2 + 1$

c Linear. $y = 3x + 5$

d Exponential. $y = 5^x$

e Quadratic. $y = x(x + 1)$

f Exponential. $y = 10^x$

g Exponential. $y = 4 \times 2^x$, i.e. $y = 2^{x+2}$

h Reciprocal. $y = \frac{-24}{x}$

i Cubic. $y = 2x(x + 3)(x - 3)$

8 The other two angles of the triangle are of size 60° and 110°.

9 $T_1 = 0.8$, $T_{n+1} = 5 \times T_n$.

10 **a** 8×10^{11} **b** 8×10^{11} **c** 8×10^{21}

d 1.6×10^9 **e** 2×10^{-3} **f** 5×10^2

11 **a** Sequence 1: 5, 17, 53, 161, 485.

Sequence 2: 0.125, 0.25, 0.5, 1, 2.

Sequence 3: −5, 5, 15, 25, 35.

b Sequence 1: Neither.

Sequence 2: Geometric.

Sequence 3: Arithmetic.

c Sequence 1: 721

Sequence 2: 3.875

Sequence 3: 75.

d Sequence 1: 774 840 977

Sequence 2: 16 384

Sequence 3: 165.

e Sequence 1: 1 162 261 446

Sequence 2: 32 767.875

Sequence 3: 1440.

12 **a** $y = 5x - 1$

b $y = 23x - 29$ is tangent at (2, 17) and $y = 23x + 35$ is tangent at (−2, −11).

13 **a** The display tells us that from $x = 1$ to $x = 6$ the function has an average rate of change of 64.

b The display tells us that at $x = 5$ the function has an instantaneous rate of change of 105.

I.e., at $x = 5$, $\frac{df}{dx} = 105$.

14 C

15 **a** 3 **b** positive **c** negative

16 Required x-coordinates are 4 and 6. Required y-coordinate is 2.816.

Exercise 6A PAGE 117

1 $10r + 3$ **2** $3 + 6k - 18k^2$

3 $15r^2 - 2r + 15$ **4** $8p^3 + 9p^2 - 14$

5 $36t^2 + 18t - 8$

6 **a** 16 **b** 26 **c** 36

7 **a** 12 **b** 18 **c** −24

8 **a** 20π **b** 6π **c** 140

9 **a** 32π **b** 48π **c** 60π

10 **a** 4π **b** 36π **c** 400π

11 **a** $\frac{4\pi t^2}{25}$ m^2 **b** 0.64π m^2

c $\frac{8\pi t}{25}$ m^2/s **d** 0.96π m^2/s

12 **a** 120 **b** 3870

c 750 bct/h **d** $(500 + 30t^2)$ bct/h

e **i** 620 bct/h **ii** 1250 bct/h

iii 3500 bct/h

13 **a** 400 **b** 50 units/h **c** 8

d **i** 57 units/h **ii** 66 units/h

iii 69 units/h

14 **a** **i** 0.2 L **ii** 2088 L

b **i** 0.03 L/min **ii** 0.25 L/min

iii 2.89 L/min

15 **a** **i** 42 **ii** 44

iii 47 **iv** 70

b $(0.2t + 2)$ deer/yr

c **i** 3 deer/yr **ii** 4 deer/yr

iii 6 deer/yr

16 **a** 150 000 tonnes

b 48 000 tonnes

c Rate of decrease = $8000 + 840t - 60t^2$,

i.e. $\frac{dT}{dt} = (60t^2 - 840t - 8000)$ tonnes/year

d **i** 9440 tonnes/year **ii** 10 400 tonnes/year

iii 10 940 tonnes/year

ISBN 9780170390408

17 a 1000 cm^3 **b** 992.4 cm^3

c $(0.2t - 4)$ cm^3/s

d i -4 cm^3/s **ii** -3.4 cm^3/s

e 20 seconds **f** $a = 0, b = 20$.

Exercise 6B PAGE 126

(Sketches not given here – check using a graphic calculator).

1 c $(-3, 61)$

2 c $(3, -50.5)$

3 Local maximum point at $(-3, 20)$.
Local minimum point at $(1, -12)$.

4 Local maximum point at $(1, 37)$.
Local minimum point at $(5, 5)$.

5 Local (and global) maximum point at $(2, 9)$.

6 Horizontal inflection at $(0, 0)$.

7 Local (and global) minimum point at $(0, 0)$.

8 Local minimum point at $(0, 0)$.
Local maximum point at $(2, 4)$.

9 Local (and global) minimum point at $(1, 5)$.

10 Local (and global) min at $(-2, -22)$.
Local max at $(0, 10)$. Local min at $(1, 5)$.

11 a $(0, 0)$

b $(-3, 0)$, $(0, 0)$

c As $x \to +\infty, y \to +\infty$. As $x \to -\infty, y \to -\infty$.

d Local maximum point at $(-3, 0)$.
Local minimum point at $(-1, -4)$.

e Use a graphic calculator to check validity of your sketch.

f Minimum value is -20. Maximum value is 16.

12 Local maximum point at $(0, 0)$.
Local minimum point at $(1, -1)$.

a -1 **b** -5

Exercise 6C PAGE 130

Each solution should clearly show the use of calculus and justify a maximum (or minimum) value.

1 When $t = 8$, X has a local minimum value of 16.

2 When $p = 10$, A has a local maximum value of 300.

3 The maximum value of A is 20 and it occurs when $x = 10$ and $y = 2$.

4 The maximum value of A is 13.5 and it occurs when $x = 4.5$ and $y = 3$.

5 When $x = 35$ the maximum profit of \$725 is realised.

6 When $x = 120$ the maximum profit of \$9400 is realised.

7 a 25 m × 25 m

b 50 m × 25 m (with the existing wall forming one of the 50 m sides).

8 The manufacturer should spend \$30 000 on advertising to achieve the maximum profit of \$140 000.

9 For maximium capacity the dimensions need to be width 0.4m, length 0.6 m, height 0.5 m.

10 The maximum capacity is achieved when $x = 10$.

11 6 cm should be turned up along each edge to maximise the capacity of the gutter.

12 a $\$(2500 + 500x - 25x^2)$

b 10 **c** \$2

d 5000 **e** \$5000

13 Minimum N is 1100, to the nearest 100. (When $t = 16$).
Maximum N is 4600, to the nearest 100. (When $t = 24$).

14 a The body is 105 m from the origin after three seconds.

b $(t^2 - 12t + 50)$ m/s

c The initial velocity of the body is 50 m/s.

d When $t = 6$ the body is moving with minimum velocity and the body is then 156 metres from the origin.

15 81 mm, 880 cm

16 The owner should spend \$12 500 on security.

Exercise 6D PAGE 134

1 $6r - \dfrac{5}{r^2}$

2 a 100 **b** 40 **c** 20

3 Maximum at $(-\sqrt{2}, -2\sqrt{2})$, minimum at $(\sqrt{2}, 2\sqrt{2})$.

4 Minimum at $(-2, 9)$, maximum at $(2, 1)$.

5 Maximum at $(-4, -18)$.

6 a $\dfrac{500}{x^2}$ **b** $\left(x^2 + \dfrac{2000}{x}\right)$ cm^2

c 10, 5, 300 cm^2

7 Correct to one decimal place, the base radius needs to be 4.4 cm and the height 8.8 cm.

8 When the base radius is 3.5 cm (correct to one decimal place) and the height is 14.0 cm (correct to one decimal place) the cost of material is minimised.

Miscellaneous exercise six PAGE 135

1 a 5^2 **b** 5^4 **c** 5^3 **d** 5^0

e 5^3 **f** 5^6 **g** 5^5 **h** 5^4

i 5^7 **j** 5^3 **k** 5^2 **l** 5^1

m 5^{10} **n** 5^4 **o** 5^{17} **p** 5^2

ISBN 9780170390408

q 5^6 **r** 5^3 **s** 5^5 **t** 5^6
u 5^7 **v** 5^8 **w** 5^2 **x** 5^3
y 5^2 **z** 5^2

2 **a** a^4 **b** $\frac{5a^7}{b^8}$ **c** $2+2^n$
d $x+2x^4$ **e** 2^x **f** $\frac{3}{5}$

3 The first six terms are 97, 108, 119, 130, 141, 152.

4 The first five terms are 350, 70, 14, 2.8, 0.56.

5 **a** 6 **b** 3 **c** 9

6 **a** 60 **b** 24 **c** 105

7 **a** 15 **b** **i** 55 **ii** 120

8 **a** 14535 **b** −442 860 **c** 12 582 906
d 500 **e** 5

9 From the graph, some of the points that the tangent at $x = 1$ seems to pass through are (−3, −11), (0, −2), (1, 1) and (3, 7). Thus the tangent at $x = 1$ has a gradient of 3. Thus the gradient of $f(x)$ at $x = 1$ is 3.
From the graph, some of the points that the tangent at $x = 2$ seems to pass through are (0, −16), (2, 8) and (3, 20). Thus the tangent at $x = 2$ has a gradient of 12. Thus the gradient of $f(x)$ at $x = 2$ is 12.
Finding the gradient of $y = x^3$ at $x = 1$ and at $x = 2$ using calculus confirms these values.

10 2

11 **a** 10 **b** 3 **c** 13

12 **a** D, H, K, P **b** B, F, I, K, N, O
c G, H, L, M **d** A, C, D, E, J, P

13 At the point (1, 2).

14 **a** (−1, −3), (1, 3) **b** (0.25, −1.25)

15 **a** 343 062, 1 698 992, 5 308 522
b 67 513, 223 973, 526 233

16 $y = 25x + 185$ (at the point (−5, 60)) and
$y = 25x - 71$ (at the point (3, 4)).

17 **a** $a = -20, c = 260$ **b** $\$(260p - 20p^2)$
c $\$(400p - 20p^2 - 1820)$ **d** 10, 60, \$180

18 **a** 1.4×10^{24} **b** 7.2×10^{12}
c 6.8×10^{12} **d** 3.5×10^1
e 7×10^{24} **f** 2.45×10^{14}

19 **a** 200 **b** 1500 **c** 130
d **i** 30 organisms/h **ii** 105 organisms/h
iii 330 organisms/h

20 **a** The display tells us that from $x = 1$ to $x = 3$ the given function has an average rate of change of 41.
b The display tells us that at $x = 3$ the given function has an instantaneous rate of change of 109,
i.e., at $x = 3$, $\frac{df}{dx} = 109$.

21 **a** $\$(10 - 0.2x)$ per metre
b $(500 + 25x)$ metres
c $\$(5000 + 150x - 5x^2)$
d 15. Negative coefficient of x^2 in the quadratic, hence turning point is a maximum.

Exercise 7A PAGE 144

1 $\frac{1}{8}x^8 + c$ **2** $\frac{1}{6}x^6 + c$ **3** $\frac{1}{5}x^5 + c$

4 $\frac{1}{4}x^4 + c$ **5** $\frac{1}{3}x^3 + c$ **6** $\frac{1}{2}x^2 + c$

7 $x + c$ **8** $4x^3 + c$ **9** $2x^6 + c$

10 $2x^4 + c$ **11** $7x^2 + c$ **12** $6x + c$

13 $x^3 + 3x^2 + c$ **14** $2x^3 - x + c$

15 $7x + 3x^4 + c$ **16** $3x^2 - 3x^5 + c$

17 $7x - 4x^2 + c$ **18** $\frac{x^3}{3} + 3x + c$

19 $3x^6 + x + c$ **20** $2x^3 + \frac{1}{2}x^2 + c$

21 $4x^3 + 2x^4 + 2x + c$ **22** $x^3 - x^2 + \frac{1}{6}x^6 + c$

23 $x + \frac{1}{2}x^2 + \frac{1}{3}x^3 + c$ **24** $3x^4 + 3x^2 + 5x + c$

25 $x^3 + 5x^2 + 8x + c$ **26** $3x^3 + 4x^2 - x + c$

27 $\frac{1}{3}x^3 - 4x + c$ **28** $\frac{1}{3}x^3 - x^2 - 3x + c$

29 $2x^4 + x^3 + c$ **30** $x^4 + 4x^3 + 2x^2 + c$

31 $y = 2x^3 + 7$ **32** $y = \frac{3}{2}x^2 + 2x - 2$

33 $y = x^3 - x^2 + 6$ **34** $y = 2x^3 - 5x + 3$

35 $y = 3x + 2x^4 + 7$

36 **a** $f(x) = \frac{1}{2}x^3 + 2x^2 - x - 2$ **b** 8

37 **a** $\frac{3}{2}x^2 - 6x + 6$ **b** 24
c $a = -4$ or 8

38 $p = 27$ **39** (−4, 0), (0, 0), (4, 0)

40 $k = -1$

ISBN 9780170390408

Exercise 7B PAGE 146

1 $V = 3t^2 + 5t + 30$

2 **a** $x = t^2 - 6t + 7$ **b** 23 **c** 1 or 5

3 **a** $A = 2r^2 + 3r^4 + 2$ **b** 58

4 **a** $C = x^2 + 3x + 100$ **b** $C = x^3 + x^2 + 5000$

5 **a** $R = 50x$ **b** $R = 50x - 0.025x^2$

6 \$38 000 **7** $(7000 - 20t - 5t^2)$ cm^3

8 Increasing. $A = 100t + 10\,000$

9 $C = 40x + 1000$

10 $R = 200x - \frac{1}{20}x^2$, 150 000

11 **a** 29 **b** 43 **c** 176

Exercise 7C PAGE 150

1 $\frac{x^3}{3} + c$ **2** $\frac{x^2}{2} + c$ **3** $\frac{x^4}{4} + c$

4 $2x + c$ **5** $2x^5 + c$ **6** $2x^4 + c$

7 $2x^2 + x + c$ **8** $2x^3 - 5x + c$ **9** $4x^2 - 7x + c$

10 $\frac{x^2}{2} + 3x^3 + c$ **11** $\frac{x^2}{2} - x + c$

12 $2x^3 + \frac{11x^2}{2} + 3x + c$ **13** $2x^3 + 3x^2 + c$

14 $2x^4 - x^3 + c$ **15** $\frac{3x^4}{2} + 4x^3 + 3x^2 + c$

Miscellaneous exercise seven PAGE 150

1 10^4 **2** 10^{-1} **3** 10^6

4 10^8 **5** 10^2 **6** 10^6

7 $10^{0.5}$ **8** $10^{\frac{1}{3}}$ **9** $10^{1.5}$

10 0 **11** 5 **12** 5

13 4 **14** 5 **15** 4

16 5 **17** 8 **18** 2

19 $T_1 = 10, T_{n+1} = T_n + 6$.
The sum of the first fifteen terms exceeds the fifteenth term by 686 (i.e. by the sum of the first 14 terms).

20 **a** 0 **b** 5 **c** $10x + 5$

d $15x^2 + 10x + 5$ **e** $2x + \frac{1}{2\sqrt{x}}$ **f** $-\frac{3}{2x^4}$

21 **a** 29 **b** 9 **c** $8x - 3$ **d** 21

22 8 **23** 6

24 $(-2, 3000), (16, 84)$

25 **a** $y = 5^x + 1$ is $y = 5^x$ with 1 added to the right-hand side.
Thus the graph of $y = 5^x + 1$ is that of $y = 5^x$ translated up one unit.

b $y = 5^{x+1}$ is $y = 5^x$ with the x replaced by $x + 1$.
Thus the graph of $y = 5^{x+1}$ is that of $y = 5^x$ moved left 1 unit.
Alternatively we could write $y = 5^{x+1}$ as $y = 5 \times 5^x$ which is $y = 5^x$ with the right-hand side multiplied by 5. Thus the graph of $y = 5^{x+1}$ is also that of $y = 5^x$ dilated parallel to the y-axis, scale factor 5.

c $y = 5^{-x}$ is $y = 5^x$ with the x replaced by $-x$.
Thus the graph of $y = 5^{-x}$ is that of $y = 5^x$ reflected in the y-axis.

d Writing $y = \frac{1}{5^x}$ as $y = 5^{-x}$ we see that the answers to this part will be as for part **c**.

26 9

27 **a** $(-1, -2)$ **b** $(-1, -3), (1, 3)$

28 **a** $a = 7, b = 3$ **b** $(0, -21)$

c Gradient is -10 at $(-3, 0)$. Gradient is 10 at $(7, 0)$.

d $(5, -16)$ **e** $y = -4x - 21$

29 $a = 3, b = 4$. Gradient at P is -7. Gradient at Q is 7. Gradient at R is 1.

30 **a** $(2000 - 40x + 0.2x^2)$ dollars per unit

b \$500 per unit

c $P(x) = 500x - 20\,000 + 20x^2 - \frac{x^3}{15}$ dollars

d \$2000 per unit

31 From the display we can conclude that the graph of $y = 4x^3 + 9x^2 - 210x + 75$ has two stationary points, one is at $(-5, 850)$ and the other is at $(3.5, -378.25)$.

32 **a**

Width (cm)	Length (cm)	Height (cm)	Volume (cm^3)
10	20	120	24 000
20	40	90	72 000
30	60	60	108 000
40	80	30	96 000

A continuation of the table, for suitably chosen values for the width, leads to maximum volume achieved when, to the nearest cm, the width is 33 cm, the length is 66 cm and the height is 51 cm.

b Volume $= 300w^2 - 6w^3$.
Calculus, and consideration of the graph of $f(w) = 300w^2 - 6w^3$, confirms that volume is maximised for $w = \frac{100}{3}$ cm, i.e. 33 cm (nearest cm).

33 Base 4 m, height 8 m, area 32m^2.

ISBN 9780170390408

Exercise 8A PAGE 159

1 (Graph not shown here.)

a The car reaches C at 11.54 a.m. and the truck reaches town C at 12.15 p.m.

b From 8.30 a.m. to 9.30 a.m. the truck maintained a steady speed of 100 km/h.

c The average speed of the truck from A to B was 87 km/h (to the nearest km/h).

d The car passes the truck at 10.30 a.m. in town B, just as the truck is about to leave B.

2

A	2 m	2 m	6 m/s	6 m/s
B	6 m	6 m	0 m/s	0 m/s
C	8 m	8 m	2 m/s	−2 m/s
D	5 m	−5 m	5 m/s	5 m/s
E	9 m	−9 m	3 m/s	−3 m/s
F	1 m	−1 m	7 m/s	−7 m/s

3 a 4 m/s **b** 31 m/s

4 a 6 m **b** 5 m/s **c** 23 m/s

5 a 0 m **b** −3 m/s **c** 3 m/s

6 a 0 m **b** 1 m/s **c** 37 m/s

7 a 3 m **b** 6 m/s **c** 6 m/s

8 a −3 m **b** −20 m/s **c** 8 m/s

9 a 1 m **b** −6 m/s **c** 150 m/s

10 17 m, 14 m/s

11 2 m, 4 m/s

12 a 8 m **b** 8 m/s **c** 7

13 a 6 m **b** −1 m/s **c** 4

14 a 12 m **b** 8 m/s **c** 3

15 153 m

16 a 45 m **b** 105 m **c** 10 **d** 60 m **e** 15 m

17 a 8 m **b** 20 m **c** 4 **d** 20 m **e** 3 m

18 a 40 m **b** 52 m **c** 6 **d** 20 m **e** 7 m

19 a 94 m **b** 148 m **c** 5 **d** 162 m **e** 26 m

20 a 10, 0

b 120 m, 10 m/s upwards

c 0 m/s, 125 m

21 60 m/s

22 a 12 **b** 318 m

c *A* and *B* collide 'head-on'.

Exercise 8B PAGE 163

1 23 metres

2 34 metres

3 The body is at the origin when $t = 0.6$ and when $t = 4$.

4 −5 metres

5 At the origin the velocity of the body is −6 m/s, when $t = 2$, and 6 m/s, when $t = 4$.

6 a When $t = 5$. **b** −12.5 metres

7 19 metres

8 0.5 metres

Miscellaneous exercise eight PAGE 163

1 a (0, −6) **b** (0, 15) **c** (0, 6) **d** (0, 3) **e** (0, −4) **f** (0, 12)

2 a (3, 0) **b** (−3, 0)

c (−3, 0), (3, 0) **d** (2, 0), (7, 0)

e (3, 0), (−2, 0), (3.5, 0)

f (1, 0), (−1, 0), (3, 0), (−4, 0)

3 2^{15} **4** 2^{-4} **5** 2^{-6}

6 2^{-1} **7** 2^{-2} **8** 2^{10}

9 2^{13} **10** 2^{6} **11** 2^{6}

12 2^{3} **13** 2^{0} **14** 2^{8}

15 $n = -5$ **16** $n = 5$ **17** $n = 13$

18 $n = 6$ **19** $n = 10$ **20** $n = \frac{5}{6}$

21 $n = 3$ **22** $n = 1$ **23** $n = 2$

24 15, 21, 27, 33, 39.

25 100, 93, 86, 79, 72.

26 4, 20, 100, 500, 2500.

27 6, 24, 96, 384, 1536.

28 213, 219, 225, 231, 237.

29 256, 384, 576, 864, 1296.

31 a $6x$ **b** $15x^2$ **c** $x + 3$ **d** $30x - 1$ **e** $6 + 18x$ **f** $-2x$

32 a $14x$ **b** $6x^2$

c $12x^3 + 3x^2 - 10x + 9$ **d** $9x^2 - 4x + 3$

33 7

34 a 10 **b** 6

c $6x + 2$ **d** 14

35 a (0.5, 1.25) **b** (−2, −40), (2, 40)

c (2, 10) **d** (−2, −20), (4, 16)

36 a −3 **b** −0.375

37 At the point (1, −4).

38 $y = -2x - 1$

ISBN 9780170390408

39 **a** From $t = 0$ to $t = 5$ the object travels 3 metres. The average speed is 0.6 m/sec.

b From $t = 5$ to $t = 8$ the object travels 12 metres. The average speed is 4 m/sec.

c When $t = 5$, the speed of the object is 1 m/sec.

d When $t = 8$, the speed of the object is 3 m/sec.

40 **a** –12 m/s **b** 12 m/s

c 2.5 **d** 1.5, 2.5

41 $a = 3, b = 5$. $(-1, -4)$

42 $y = 1.25x - 2$

43 $f(x) = 7x^3 - 15x^2, f(3) = 54, f'(2) = 24$

44 **a** 3399, 9744, 235 056

b 859, 1729, 14 401

45 **a** $60x + c$ **b** $30x^2 + c$

c $20x^3 + c$ **d** $15x^4 + c$

e $12x^5 + c$ **f** $10x^6 + c$

g $2x^4 - 5x^3 + 2x + c$

h $4x - \frac{3}{2}x^2 + \frac{2}{3}x^3 - \frac{1}{4}x^4 + c$

i $\frac{1}{3}x^3 - 9x + c$ **j** $12x^4 - 8x^3 + c$

46 **a** $y = 2x^2 - 3x + 3$

b $y = 2x^3 - x^2 + 4x + 7$

c $y = 2x^4 - 4x^3 - 2x^2 + 11x - 10$

47 When $x = 2, y = 64$.

48 **a** 7500 cm^3

b 29 100 cm^3

c $(5400 - 900t + \frac{75t^2}{2})$ cm^3/s

d 3750 cm^3/s, 2400 cm^3/s, 150 cm^3/s.

49 **a** y changes by 320 units (from 1 to 321) when x changes from $x = 0$ to $x = 10$.

b y changes at an average rate of 32 units per unit change in x when x changes from 0 to 10.

c When $x = 0$, the instantaneous change in y is 2 units per unit change in x.

d When $x = 10$, the instantaneous change in y is 62 units per unit change in x.

50 D

51 \$18, 15 000

52 **a** 5 metres

b 57 metres

c From 5 metres to 57 metres in 4 seconds is an average rate of change of 13 m/sec.

d When $t = 2$, the instantaneous rate of change of x is 3 m/sec.

e When $t = 6$, the instantaneous rate of change of x is 27 m/sec.

53 **a** $250 - \pi r$ **b** $A = 500r - \pi r^2$

54 \$22 003.73

55 **a** $120 - x$ **b** $\$(360 + 3x - 0.05x^2)$

c \$4.50 **d** \$405

e $P(x) = \$(120 + 5x - 0.05x^2)$

f \$5.50 **g** \$245

56 **a** 25 m **b** 5 **c** 20 m

57 $a = -0.02, b = 4.8, c = 0.018, d = -6.6, e = 50, f = 30,$ $g = 20, h = \frac{550}{3}$

INDEX

D

F

ISBN 9780170390408

ISBN 9780170390408